GASOLINE
TREASURES

WITH VALUES

MIKE BRUNER

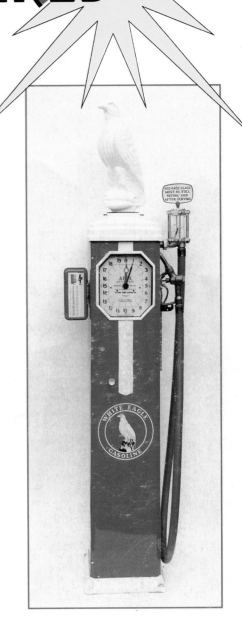

Schiffer Publishing Ltd

77 Lower Valley Road, Atglen, PA 19310

DEDICATION

To a very special person that gave everything -
and asked for nothing in return.

DR. THOMAS J. PETINGA, SR., M.D.
Doctor, Father, Scholar, Humanitarian, Friend

January 11, 1912 - December 1, 1994

Requiescat in pace

Printed in Hong Kong
ISBN: 0-7643-0050-4

Library of Congress Cataloging-in-Publication Data

Bruner, Michael.
    Gasoline treasures / Michael Bruner.
      p. cm.
    ISBN 0-7643-0050-4 (paper)
    1. Service stations--Collectibles--United States--
Catalogs. 2. Petroleum industry and trade--Col-
lectibles--United States--Catalogs. 3. Americana--
Collectors and collecting--Catalogs. I. Title.
NK808.B78   1996
629.28'6'0973075--dc20   96-4676
                                          CIP

Published by Schiffer Publishing, Ltd.
77 Lower Valley Road
Atglen, PA 19310
Please write for a free catalog.
This book may be purchased from the publisher.
Please include $2.95 postage.
Try your bookstore first.

We are interested in hearing from authors
with book ideas on related subjects.

SPECIAL THANKS

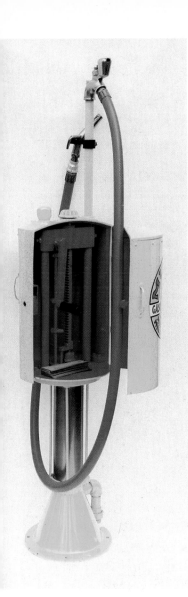

This book has been made possible through the efforts of two dedicated collectors and their wives:

GENE and RED SONNEN
&
TOM and SUSAN DAHL

The entire contents of photographs found in this book is taken from their fantastic collections, and I am indebted to Gene Sonnen for his hard work to bring this project together. Thank you for allowing us to share your collections through the pages of this book.

ENJOY!

ACKNOWLEDGEMENTS

Much of the credit for a completed book should be given to the thoughtful people that helped *"behind the scenes."* I'd like to take a moment and acknowledge certain key people that helped with their special talents and encouragement to bring together this book.

To Peter Schiffer, my publisher, and Douglas Congdon-Martin, my editor, my sincere appreciation for your support throughout this project.

To Sharon Callender and the crew at Drayton Printing and Copy Center, for your assistance with text layout.

To John R. Heafield at PHOTOFAST, Birmingham, Michigan, whose photofinishing is used in the majority of my photographic work.

The outstanding magazine ads placed throughout the book are reprints from the National Petroleum News in the late 1920s and are courtesy of two very special friends from Tulsa, Oklahoma, Dave and Kathy Lane.

My sincere appreciation
Mike Bruner

ISLANDS OF THE ROAD

IN THE BEGINNING were motorcars
dashing about, throwing oil and grease
and backfiring sparks like shooting stars
not giving the countryside a moment's notice.

Even a short trip could take all day,
as the only pavement was good intentions.
Stops were quick. They needed a faster way
to fill with fuel these thirsty inventions.

And so along came the filling station,
each one a slightly more curious design.
By the thousands they sprung up across the nation
with their shining islands hosting porcelain signs.

On top of the tall gas pumps would nestle
a lighted glass globe with curious script.
New words like no-knock, benzol and ethyl
came to be used in the course of a trip.

Attendants in bow ties and clean jump suits
sprang to the call to check oil and gas
like a racer's pit stop. Then off with a toot,
there were miles to go and buggies to pass.

The green dinosaur, the big yellow shell,
the flying red horse and the Texaco star,
all stood tall like sentinels to tell
of rest and refill for truck or car.

Now the old station's a fashion shop,
the pumps are novelties that sit in the den,
the globes have gone to the auction block,
while self-serve replaced the bow-tied men.

So I jump in my import, take the backroads of time,
just hoping the station's around the next curve,
with someone to greet me with a word that's kind.
"Check your oil, sir? We're here to serve!"

Thomas Kleeberg

CONTENTS

INTRODUCTION

We would be hard pressed to think of anything that occupies more of our time and effort than the automobile. There is the telephone, of course, but given our ability to relate to the auto, the telephone would have to take second place. Few would debate that the need for quick reliable transportation has brought about the largest industry in the world, the automobile.

Because of the central place of the automobile in American life, the artifacts that surround it are in a very real way a part of Americana. In the automobile's relatively short history, the service station is responsible for the millions of items that today are much sought after collectibles. This book is presented as a tribute to those stations.

As you go through the book, there will be very few things that can still be found in service today.

Technology and progress have taken their toll, even on the American gas station. The products and services of the gas stations were constantly in the public eye, and few petroleum companies would allow their stations to stagnate due to being behind the times. Newer was better, and nothing was spared.

The haste to constantly change our image has made us into what has been called a "disposable" society. While many bemoan this fact, it has proven to be a boon for today's collectors. We seem to thrive on what people have replaced and rejected. Fortunately, there seems to be no end in sight. We are still the happy recipients of a never-ending array of castaway advertising products.

So, sit back and enjoy this brief view of gas station memories - you'll be on a trip through American service station history.

GASOLINE PUMP GLOBES

The years 1910-25 saw the rise of gasoline marketing as big business. As a way to catch the public's attention, almost every retail station utilized globes on top of their pumps to promote gasoline sales. The earliest globes were of solid one-piece construction. They had fairly simple graphics that were etched into the glass most of the time.

Metal band globes were also made in the early era. Their sturdy construction with interchangeable lenses was a plus to the local retailer. They continued to be manufactured until around 1930, at which time all-glass globes came back into vogue.

No matter how appealing a globe was in daylight, it was always night time that revealed the true beauty of these pump-top sentinels. This, coupled with the high cost of metal band bodies, may explain why manufacturers were going back to all-glass globes.

The basic all-glass globe of this era will have three-piece construction, consisting of a body and two lenses. Several styles can be found, the most common being a "solid" glass face, with lenses that simply fit over the front and back. They were attached with two small bolts with leather washers on each lens.

A variation of this is the "Hull" body that had a hollow face with a .5" lip of glass around its perimeter to facilitate attaching the lenses.

Another variation used a metal band to hold each lens in place. These are known as "Gill" bodies. A beautiful variation of these is found in the "Ripple" body style. They were made in several colors, and their textured glass surface brought a new dimension to their appearance.

As the years passed, the rising costs involved in glass globe production forced manufacturers to produce globes made of plastic. These were mostly manufactured in the era after 1950.

Due to the changing design of American service stations, new pumps eliminated globes altogether. By 1970, only a few die-hard stations still utilized globes, a tribute to the thousands of globes before them that have become a memory of roadside America.

This 16" low profile metal band Pro-To-Co globe was found in a Chicago area antique show. The use of the words *"Tank Car"* gave emphasis to the fact that the gasoline was brought from the producer to the local consumer via rail car. No doubt the abbreviated *"Pro-To-Co"* stands for *"Producer-To Consumer"*. $750

This beautiful red Powerflash ripple body globe was from the H.K. Stahl Company of St. Paul, Minnesota. Founded in 1888, this company only recently went out of business. Not only did they market Powerflash gasoline, their Trophy Motor Oil was a well-known product trademark for years. $1,700

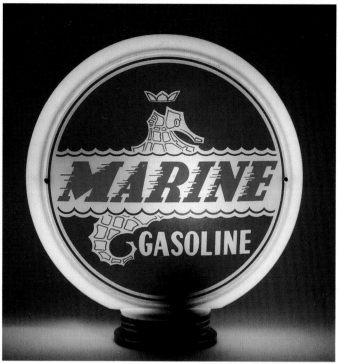

Foster Oil Company used this graphic Supertane 13.5" globe in the 1930s. It incorporates the use of the later style Ethyl burst trademark. $450

Many petroleum companies took advantage of the large potential in the Marine gasoline market. This unusual 13.5" glass body Marine Gasoline globe dates from the era around the 1930s. $550

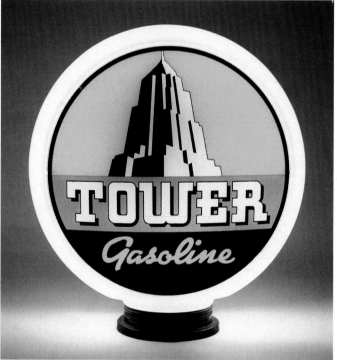

This rare 12" metal band globe from the Boner Oil Company represents one of the earliest known globes. Its lenses were not silk-screened as normally found, but were actually hand stenciled. The metal band body used a chimney-type system inside to dissipate the heat so that the paint would not deteriorate or the glass would not crack. The Boner Oil Company was located in the Beloit, Wisconsin, area. $1,000

The 1930s saw the use of this Tower Gasoline 13.5" globe. $650

Southern Oil Stores Inc. were the local retailers for the Dixie Vim brand regular gasoline featured on this yellow ripple body globe. It dates from the 1950s and has the popular slogan *"From Tank Car to Car Tank."* $1,850

Similar in design to the Southern Oil Stores globe, this Dixie Vim green ripple body globe utilized the later style Ethyl burst trademark. In case the trademark was overlooked, the word *"Ethyl"* in large, red letters made a standout appearance. $2,700

400 Aviation Dry Gasoline was produced by the Shell Oil Company of California in the 1930s. This example being found on a 15" high profile metal band globe. $2,700

Here's a closeup of the embossing on the bottom of the 400 Aviation globe frame.

This 13.5" Red Indian globe was used by a Canadian oil company. $1,000

Old Trusty Gasoline and Motor Oils is featured on this 13.5" white ripple body globe. $1,200

White Rose Gasoline used their popular trademark on this 13.5" glass body globe. It dates to the era around the 1940s. $950

Gibson Brothers Oil Company was a member of the Independent Oil Men of America and is found on this 15" high profile metal band globe dating from the 1920s. $900

This smaller-sized single-piece Sinclair Oils globe dates to before 1920. Its possible use was on oil cabinets. $850

Sinclair Gasoline used this one-piece etched globe to advertise their product. It incorporates their well-known bar logo found on advertising from their earliest years. $800

Shell Oil used this low-profile 15" metal band globe in the 1920s. It incorporates their well-known trademark which is still in use today. $2,500

This similar Shell globe utilizes a 15" high-profile metal band globe. Although the Shell logo differs slightly from the previous photo, it still dates from before 1930. $2,500

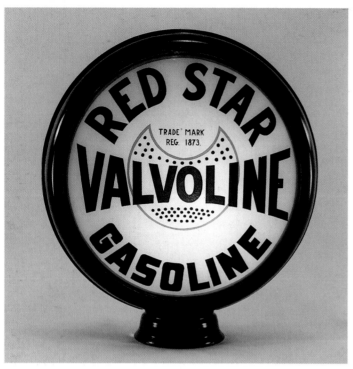

Red Star Gasoline by Valvoline appears on this high-profile 15" metal band globe. Valvoline started in 1873 as a lubrication for steam valves. The crescent shaped object outlined in red at the center of Valvoline's logo is supposed to be representative of a steam valve. $900

We-No-Nah Gasoline was produced in the 1920 era and is found on this 15" high-profile metal band globe. $450

Phillips Petroleum was the producer of this Unique 15" high-profile metal band globe. It dates from the 1920s era. $600

Super-Par "100" Gasoline is seen on this red ripple body globe with threaded copper base. The style of the aircraft seems to date this one from the era around 1950. $1,600

Cities Service utilized many styles of glass and plastic globes through the years. This cloverleaf style Koolmotor globe dates from the 1930s era and utilizes a heavy glass body. $1000

This etched one-piece body Valvoline globe dates from the 1920s. It has original paint which is quite difficult to find. $2,500

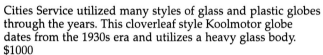

Early Cities Service Oils Aviation lenses are found on this 15" low-profile metal band globe. Many early petroleum companies avoided buying into the Ethyl Corporation's anti-knock product by marketing their own high-octane fuels. This particular globe came from O'Brien Oil Company in Superior, Wisconsin, and was one of six discovered there several years ago in an abandoned warehouse. The word *"aviation"* is hand stenciled on the lenses. $550

Super-Shell Ethyl takes the shape of their well-known trademark in this outstanding one-piece Shell globe. It is ink stamped *"April 1930"* on the inside. $950

This etched one-piece round Shell globe pre-dates the familiar clamshell style globes and is quite scarce. $700

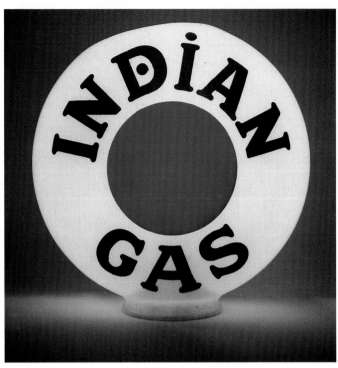

Indian Gas was bought out by Texaco in the 1930s. The one-piece globe shown here has a manufacturer's defect insofar as it is not uniformly round. $1,200

Shell Oil Company produced clamshell single-piece globes by the thousands, many of which survive today. $400

Pioneer Gasoline makes the grade on this rare 15" high-profile metal banded globe. Pioneer was a Minneapolis company that was later bought out by Skelly. The outstanding graphics featured on the lenses score big with collectors. $4,500

Phillips 66 with Ethyl is featured on this scarce 13.5" plastic globe. $450

Buffalo Gasoline was a product of Westland Petroleum of Minot, North Dakota. This outstanding three-piece narrow body globe features their running buffalo trademark. $1,000

In an effort to market their own high-performance gasoline, Phillips Petroleum introduced its 77 brand of gas in the 1930s. This particular design was followed by a Phillips 77 shield with wings and the word *"Aviation"* on it. The plastic wide-body globe that is shown is an eggshell color. $500

Airports often had their own high-octane fuels. This 13.5" narrow band globe incorporates the famous Pegasus logo of Mobil Oil Company and has the words *"Mobilgas Aircraft"* to designate its use. $900

Valvoline, Kendall, Pennzoil, and Wolf's Head were all producers of gasolines that were eventually discontinued. However, these companies continued to do big business in the oil market. This 13.5" narrow body globe has their famous Wolf's Head trademark. $700

Texaco Ethyl Gasoline was promoted with the help of this one-piece raised-letter globe. It is stenciled *"1930"* on the inside. $1,200

Station WNAX was located in Sioux Falls, South Dakota, and was owned by the Gurney Petroleum Company. This regional radio station is the theme on their three-piece 13.5" narrow body globe. $950

This Texaco one-piece raised-letter globe dates from around 1930. $850

Texaco also used this etched one-piece globe in the 1930s. It was manufactured by Solar Electric Company of Chicago. $1,500

Koolmotor was a trade name marketed by Cities Service Company. Featured here is their familiar cloverleaf design on a one-piece glass body. $1,800

Ford Benzol Gasoline was marketed by Ford Motor Company at several of the major name brand gasoline stations in the Detroit area. Its sale was regarded as an alternative petroleum to the major brand offered at the station. This globe features a keyhole design with a silk-screened paint job. $4,000

Cities Service also manufactured a limited quantity of these three-piece cloverleaf shaped globes with bolt-on "round" lenses. These date from the 1930s, and are quite desirable in the collectors' market. $1,000

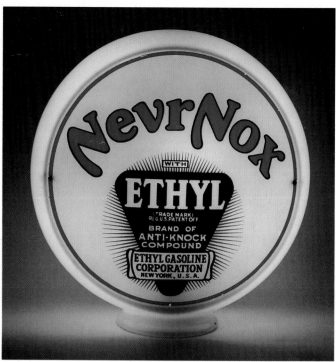

Hy-Speed Gas was located in Michigan and went out of their way to make sure the customer knew it was a local product on this one-piece etched globe. $1,200

This 13.5" three-piece narrow band globe headlines Nevr Nox Gasoline which was produced by the D.X. Corporation. Notice that the Ethyl logo is missing its normal yellow color. This deviation was no doubt intentional, possibly as a way to help to identify the product Nevr Nox. $450

Royal 400 Gasoline was marketed by the Independent Oil Men of America and distributed from Spencer, Iowa. This globe features an etched one-piece body with a stenciled Independent logo in its center that was fired on. $2,500

Although only one side of this rare 15" metal band globe is visible, there is a matching lens on its other side. Red Hat Gasoline was involved in a litigation settlement in the 1920s with Standard Oil Company which in essence forced them to discontinue the use of its Red Hat trademark because of its similarity to the Red Crown trademark of Standard Oil. $1,500

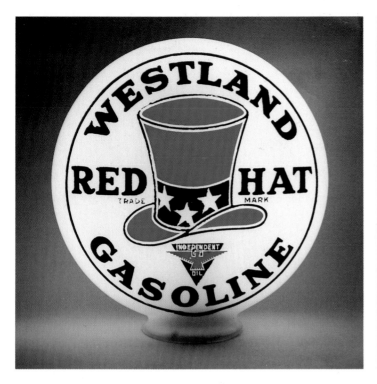

Westland Red Hat Gasoline used this etched one-piece globe in the 1920s. It is extremely rare with its original paint still intact. $4,000

This 13.5" three-piece white ripple body globe features lenses from the Kanotex Company. Their Bondified Gasoline was mostly marketed in Kansas, Oklahoma, and Texas, which when abbreviated spell the word *"Kanotex"*. The star in the center of the logo is surrounded by a sunflower. $1,500

Red Head Hi-Octane Gasoline was marketed in Ohio and Virginia and is the theme on this 13.5" three-piece globe. $450

North Star Petroleum had their unusual R-Tane marketed with the help of this 13.5" three-piece narrow body globe. $450

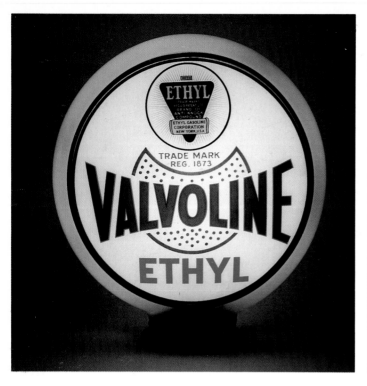

Valvoline used this gill body globe to market its Ethyl Gasoline. This was one of Valvoline's last glass body globes before the introduction of plastic bodies. $850

Sinclair Aircraft Gasoline is found on this rare three-piece glass narrow body globe. Almost all the Sinclair Aircraft globes found to date are one-piece glass bodies. $3,000

Here's a closeup of the manufacturers decal found on the top of the Sinclair Aircraft Globe. Cincinnati Advertising Products Company was one of the largest manufacturers of gasoline globes and made tens of thousands of globes in its long history.

The very rare Roxanna-styled clamshell is highly sought after by Shell Gasoline Enthusiasts. This 15" high-profile metal banded globe set a world's record at the time it was sold. The globe features two inserts which have stenciled-on graphics and the body has the very desirable jeweled style band. Restoration of this style band is extremely difficult due to the fact that the jewels have brass retainers on the inside making their removal and reinstallation a timely chore. $10,000

Here is a night shot of the Roxanna jeweled globe. What a beauty!

Deep Rock Gasoline used this high-profile metal banded globe with metal inserts in the era around 1930. $1,000

A not-so-happy lion is the trademark on this Wood Hi-Test 13.5" three-piece glass body globe. $900

The metal lenses not only had stenciled-on graphics, but also incorporated small holes through which light could pass as the photo suggests, giving the letters a jeweled effect. $1,000

Lion Gasoline used this three-piece globe from around the 1940s era. $650

Green Streak Gasoline was marketed for a limited time through Shell Oil Company in the west coast states. Shown here is a 15" metal band globe. $900

Beeline Gasoline used this yellow-bodied ripple globe to market its product in the 1940s era. $1,700

Kanotex produced this white ripple body globe in the 1940s era. $1,500

A slightly different variation was this Kanotex yellow ripple body globe featuring a 1940s-style aircraft with the letters *"Aviation"*. Again, *"Aviation"* no doubt means Kanotex's version of high-octane automobile gasoline, and was not actually intended for aircraft use. $2,800

For comparison, the next few globes have been
presented with daylight as well as night-time shots.

Torpedo Gasoline was marketed by Illinois Oil Company of
Rockford, Illinois. This 13.5" red ripple body globe incorporates the outline of their home state. $1,700

Here's a night shot of the Torpedo red ripple body globe.

Few historic Americans are as beloved or recognizable as Ben
Franklin. The outstanding graphics combined with the red
ripple body globe make this Ben Franklin Premium Regular a
winner. $2,500

Here's Ben Franklin at night.

Globe Gasoline found its way onto this blue ripple body globe which dates from the era around the 1940s. $2,800

The eerie blue glow of the ripple body globe at night almost gives the appearance of the earth's atmosphere surrounding the globe in space.

This scarce Husky Ethyl globe was distributed by Western Fuel and Oil Company in Minneapolis and is found on a gill body which dates to the 1930s. $2,500

The outstanding graphics on the Husky globe come alive when lit at night.

Sunray's eight-sided logo is highly sought after by collectors. The matching yellow ripple body globe makes this combination a winner. $2,000

Only a night shot can truly bring out the full beauty of a ripple body globe.

The Jenney Manufacturing Company of Boston used this orange ripple body to feature their Jenney Solvenized Hy-Power Gasoline. The graphics shown on the lenses depict a refinery scene that shows a bulk truck being loaded on the left, the refinery at the right, and a railroad scene with tank cars unloading the crude oil in the center. $2,800

Like other ripple body globes, this orange Jenney literally comes alive at night.

Solite Gasoline was a product distributed by Standard Oil Company and is captured on this 16" low-profile metal band globe which dates from the 1920s. $950

Farmers Union Petroleum was a St. Paul, Minnesota, based company. Their plow, rake, and hoe shield logo is featured on this yellow ripple body globe from the 1930s. $1,750

Conoco Gasoline shows only the silhouette of their well-known Minuteman trademark on this 15" high-profile metal band globe dating from the 1920's era. $2,800

Farmers Union had this 15" low-profile metal band globe in the 1920s. $1,000

The most outstanding aspect of this early 1920s Independent 15" low-profile globe is the marvelous graphics found in its trademark. $2,000

This closeup gives you an idea as to the detailed graphics found on the Independent globe.

Marathon Petroleum was one of the larger marketers of gasoline products through the years. This 13.5" three-piece narrow body globe features their well-known runner trademark with the slogan, *Best in the long run*. $600

Marathon Petroleum used this 15" low-profile metal band globe in the era around 1920. $950

Benzol Power-Miles Gas was the featured product found on this 15" high-profile metal band globe. $900

Sinclair Gasoline used this one-piece etched body globe in the era of the 1920s. H-C stood for *"High Compression"*. $750

"Quick as a rabbit" might be the slogan used by Mutual Gasoline. One thing for sure, few of these 15" low-profile metal band globes remain today. $1,400

This 13.5" three-piece narrow body glass globe marketed Skelly Aromax Gasoline and has the older style Ethyl burst logo in its center. $450

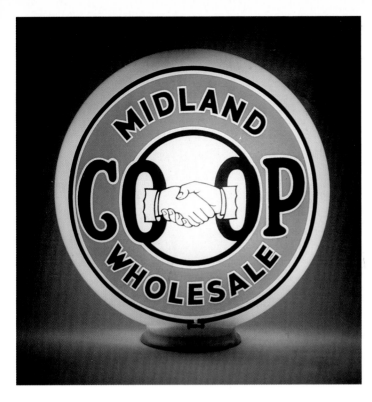

Many small cooperative petroleum companies sprang up throughout the Midwestern United States through the years. This Midland Co-Op Wholesale globe is typical of the many small-time independent retailers doing business to farmers and the motoring public around the 1930s era. $750

Any Cities Service Koolmotor globe with red letters is rare. This 15" high-profile metal band version dates from the era around 1920. $600

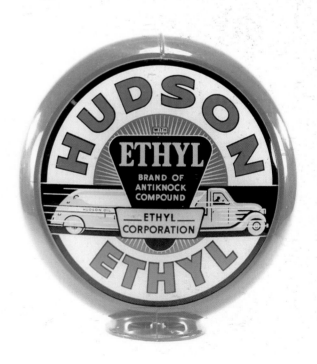

The beautiful graphics found on this Husky 13.5" gill body globe were discontinued in the mid-1940s. Rumor has it that the public did not appreciate the similarity between the sun behind the dog and the Japanese rising sun $1,500

Hudson Petroleum was one of the larger independent petroleum companies. This 13.5" two-piece plastic body features lenses that have the new style Ethyl burst logo as well as their tank truck logo trademark. $1,000

This Regular Gasoline Hudson Globe is found on an orange ripple body globe. $2,500

Bell Gasoline was marketed on this orange ripple body globe from the 1940s era. Note the oil derrick in the bell's center. $1,500

Bell Gasoline made the grade on this 13.5" red ripple body globe with the older style Ethyl burst superimposed over the Bell trademark. $1,800

D-X Petroleum Company used this plastic globe hood to emphasize the rocket power that could be found in its new Boron Gasoline in the 1950s. Very few exist today because they were manufactured of thin plastic which had a tendency to fade and develop cracks. $600

Purol Gasoline was one of only a handful of companies to utilize porcelainized metal-frame bodies for their globes. This one being a 15" high profile. $600

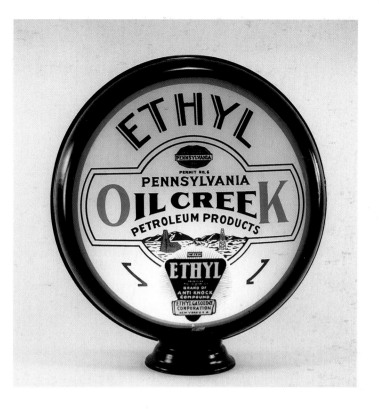

Oil Creek Petroleum Products produced this 15" high-profile metal band globe to market their Ethyl Gasoline in the 1920s. $1,200

Illinois Oil Company marketed Torpedo Gasoline in the era around 1930. This globe utilizes a 15" high-profile metal band. $850

Fyre-Drop Gasoline was an early product marketed by the Barber Oil Company of Minneapolis and Chicago. This small independent manufacturer refined their own products and this 15" low-profile metal band globe is actually painted in a two-tone paint scheme with a blue band going around its perimeter and base. $850

Barber Oil Company also distributed a much rarer Me-Tee-Or Gasoline pictured here in a 15" high-profile globe. $1,200

Pennzip Gasoline was distributed by the Pennzoil Company who today enjoy a healthy share of the motor oil market. The Pennzip Gasoline globe pictured here has a large early style Ethyl burst logo in a 13.5" three-piece narrow bodied threaded base globe. $550

Kunz Gasoline was marketed in the Minneapolis and St. Paul area beginning in the 1920s. Although its gasoline products were discontinued in the 1970s, Kunz is still in business today as an auto parts supplier. The globe pictured here is in a 15" high-profile metal band frame and is the only one known to exist. $7,500

Sunray Gasoline used this three-piece plastic body globe in the 1950s era. $750

The legendary graphics depicted on this Musgo globe set the standard for overall detail and beauty. Musgo was located in Muskegon, Michigan, and each of their globes were hand painted and fired. Hence, no two are exactly alike. As with most scarce petroleum items, the high demand for this globe has kept the available supply at a minimum, and the price at a maximum! $5,000

Only a close-up could reveal the exceptional beauty and artistry found on the chief of globes.

Utility Gasoline and Globe Gasoline with Ethyl were marketed as an inexpensive alternative to the major brands. Both globes have white ripple bodies. The one on the left having a standard base. The one on the right having a copper threaded base. $1,500 & $1,800 respectively

This variation on the Globe theme contains a gold band around its perimeter showing a silhouette of various modes of transportation of the day. Notice the biplane and early automobile designs. 1,500

Minnesota Farm Bureau was one of four gasoline co-ops in the State of Minnesota. This example being featured on a red ripple body globe. $1,700

Coastal Anti-Knock Gasoline is the brand found on this rare clear 13.5" ripple body globe. $1,800

For those into economy, Thrifty Gas fits the bill. This example dates from the 1950s and is on a three-piece red plastic body. $450

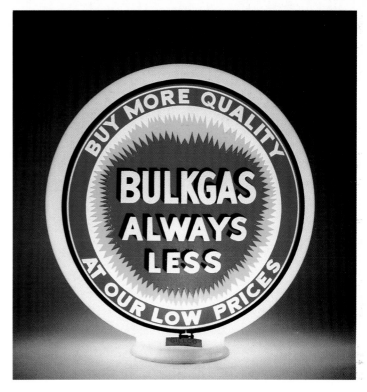

Bulkgas was an independent supplier of an economy-priced gasoline. The gill body seen here has the original Gill Company decal affixed near the base. $550

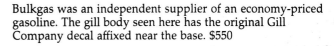

We could only guess as to the many local companies that utilized this Visible Gasoline globe. It dates to the era around 1915 and could have been used by any independent supplier. Many of the earliest pumps did not feature company-specific advertising on their globes, but were able to get by with a globe such as the one pictured here. $550

This early Red Crown Gasoline globe utilized the largest metal band frame to be found on a globe. It measures 16.5" and held two large-size lenses. $1,000

Few globes can rival the desirability of this jeweled 15" high-profile metal banded globe. Powerflash was an independent gasoline produced by H.K. Stahl of St. Paul, Minnesota. Its graphics depict the ignition of gasoline on a piston. The metal band has 24 red, white, and blue jewels which, needless to say, make this combination a standout. $4,000

Here again it takes a night shot to truly appreciate the beauty found in this exceptional metal band globe.

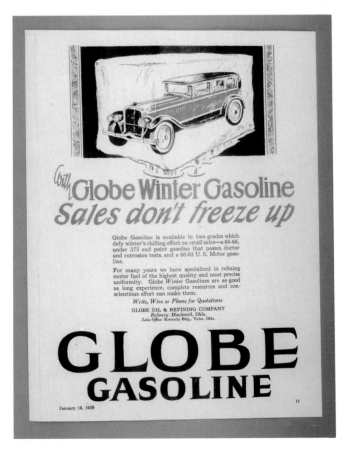

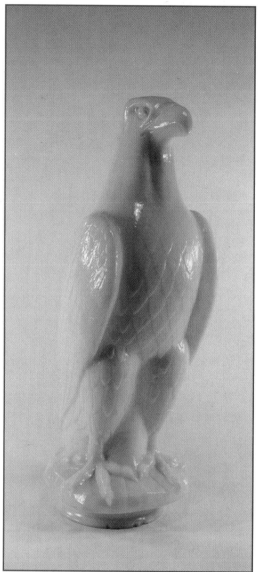

In the 1920s, White Eagle Gasoline produced three variations of one of the most beautiful figural gasoline pump globes ever manufactured. These globes were in use until the 1930s when White Eagle was bought out by Mobil. Each variation has its own distinct appearance, color, and texture, and each are given a trade name by hobby enthusiasts in the collectibles market. On the left is the most often found of the three and is called the blunt nose. This globe has the least amount of detailing work and is found in a heavy milk glass. Its appearance shows rather few feather details and a beak which makes the words *"blunt nose"* seem appropriate. Shown in the center photo, the *"full feathered"* eagle is much more scarce than the blunt nose and can be easily distinguished by the shallow base on which the eagle stands. Notice that far more detailing work was involved in the mold process. On the right is the scarcest variation, what some collectors have come to call *"the crow."* This is understandable if you look at the eagle's head. The back side of the eagle reveals a heavily embossed feathered back. All three of the White Eagle globe variations are highly sought after by collectors. $1,200, $1,700 and $1,800, respectively

The three variations are seen here standing in silent tribute to the craftsmanship represented in the art of mold making.

Even the back side of these magnificent birds presents itself as a salute to craftsmanship.

Few gasoline pump globes are as remembered as the Crown series by Standard Oil. Pictured here is Standard Oil's Premium Gasoline globe. Although difficult to tell by this night-time photo, the dark areas were painted gold. The Gold Crown Premium Gasoline was offered by Standard between 1957 and 1959. $350

Red Crown Standard was regular gasoline. These globes were made in the thousands but collector demand has not only escalated their prices, but dwindled the supply. $250

Standard Oil's early premium gasoline was Red Crown Ethyl. Notice no painting on the upper portion of the globe and that the base does not utilize an aluminum threaded ring. $750

Atlantic White Flash is presented here on this 16.5" high-profile metal band globe. Apparently Atlantic Refining was concerned that the globe by itself didn't get enough attention, as they added this marquee of porcelain on metal to the top. $850

Although these two globes appeared earlier in this chapter, they are included here so you may get an idea as to the comparative sizes of the smallest and largest metal band globes to be used.

Standard Oil's Blue Crown was designated for their economy grade gasoline, trademarked "Solite". $700

These two Sinclair globes were also featured earlier but have been included here so you may see a comparison between the largest and smallest glass body globes that were in production.

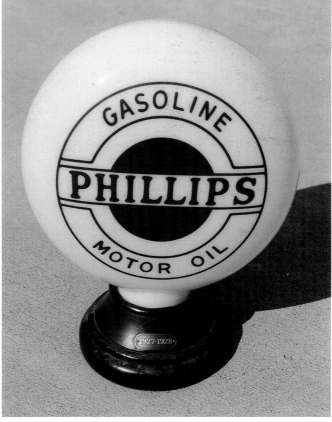

In the late 1970s Phillips 66 produced this small-size merit award for their dealers for selling high quantities of Phillips 66 Gasoline or Motor Oil. The familiar Phillips 66 shield on the front is contrasted by the early Phillips logo placed on the reverse. $450

ADVERTISING SIGNS & THERMOMETERS

Few items from filling stations are as representative of the products and services offered as the thousands of signs that were made through the years. Most of these served a particular need, such as on gasoline pumps or lubesters. Still others were found at the curbside or roadside. Even custom made signs in tin and porcelain were made, identifying an individual station or person. As you go through this chapter, look carefully at the more ornate and graphic signs. The attention to detail on some of them is nothing short of spectacular.

Skelly produced Tagolene Motor Oil as being Tailor-Made. The 30" round two-sided sign pictured here was Tailor-Made for a lollipop stand. $350

At first glance, this Texaco No Smoking strip sign appears to be the version that was made in the thousands. A closer inspection will reveal that this sign is dated March 1939 and has the desirable black "T" in the Texaco logo. $350

The king of petroleum no smoking signs might be the best way to describe this beautiful example from the Texas Company which dates to the 1920s. Its black porcelain background makes the colors vibrant. $1,500

Standard Oil produced this scarce No Smoking sign in the 1920s. $300

Even Cities Service Oils entered the No Smoking sign market with this rare strip sign. $1,200

Not to be outdone, Shell Oil also produced a No Smoking sign in their familiar red and orange color scheme. $1,500

These restroom signs are from a Shell Service Station and are made of porcelain. $225

As a marketing ploy, many service stations began an in-earnest effort to offer the cleanest, best restroom facilities to the motoring public. Part of this campaign included prominent placement of signs that would identify those dealers who felt clean and sanitary restroom facilities were important to motorists. $350

Most all early gasoline outlets did a hefty business with motor oils. The early years of marketing motor oil saw its sales from dispensing pumps called lubesters. Many companies produced their own signs identifying the particular grade of oil that was dispensed to the consumer. Some were made of porcelain, but most were made of tin. All have become highly desirable as relics of days gone by in the petroleum business. Values, clockwise from top left: pg. 44: $225, $50, $65, $150; pg. 45: $65, $275, $500, $650

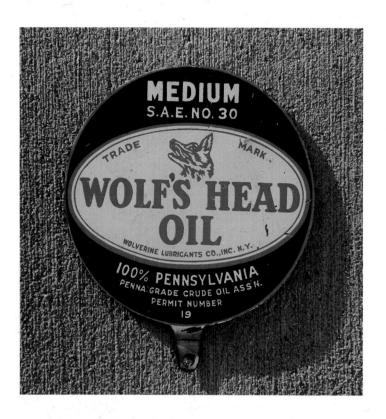

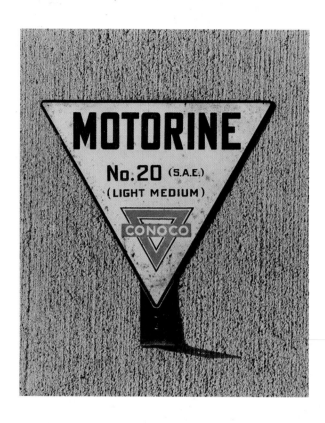

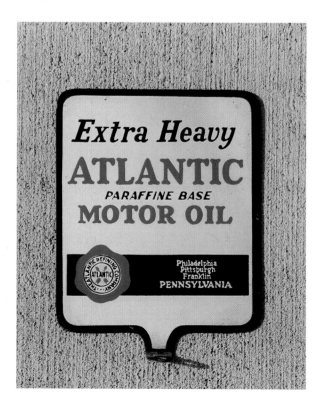

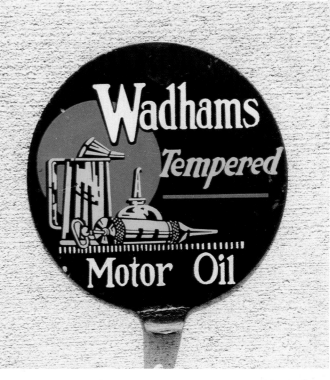

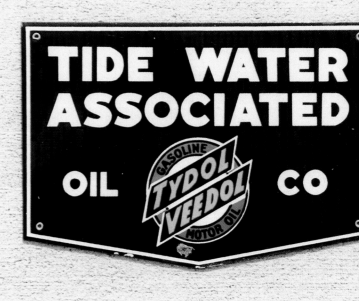

This rare oval Clipper Gasoline pump sign dates from the 1940s. $2,500

Tide Water Associated Oil Company produced this pump sign in the era around the 1940s. $750

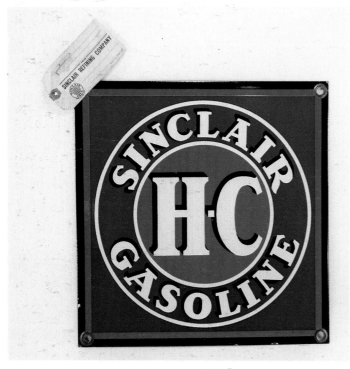

Many times petroleum bulk trucks would utilize porcelain signs prominently placed on their doors. This Sinclair H-C Gasoline sign is shown with its original shipping tag. $500

Highly desirable because of its die-cut eight-sided design and its colorful graphics, this Sunray D-X dates from the late 1950s. $750

Similar in design to Mobil Oil's shield, this Knight Oil Company Regular Gasoline pump sign dates from the 1950s. $850

Webb Petroleum was based in Minnesota. This pump sign is unusual because it is made of porcelain rather than the typical tin found on most Webb pump signs. $325

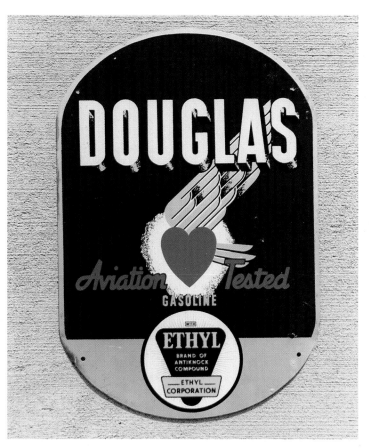

The war years saw the use of this Douglas Aviation Tested pump sign. It is unusual insofar as it is made from baked enamel on tin rather than porcelain, probably because of World War II steel shortages. $350

Outstanding graphics have put this Smith-O-Lene Gasoline pump sign on collectors' most-wanted lists. Aviation Brand was their version of high-octane fuel. $950

This Jenney Aero Solvenized pump sign dates from the 1930s. $1,500

Sterling Gasoline was one of the last fuels sold by Quaker State. This sign dates from around 1950. $350

Texhoma was a regional petroleum company which operated in the 1930s. $650

This Cosden Higher Octane pump sign dates from the 1940s. $150

Jenney Manufacturing Company of Boston features their refinery scene on this 1930s pump sign for Solvenized Hy-Power Gasoline.

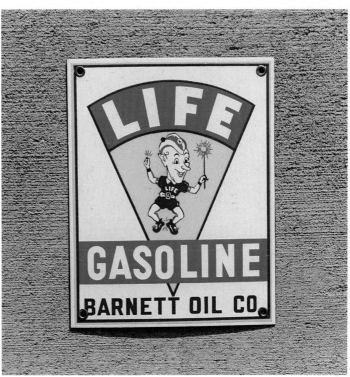

Life Gasoline's eye-catching logo is found on this 1940s porcelain pump sign. They can also be found made of tin, with identical graphics. $1,000

Signal Gasoline used this pump sign in the 1950s era. $800

These small-sized curved pump signs from Texaco depict their famous Fire-Chief and Sky Chief brands of gasoline. $150 each

All four of these pump signs are the small-sized versions. The rarest being the Marine White. $225, $150, $300 and $1,000 respectively

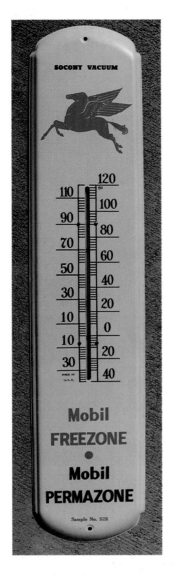

The use of porcelain on thermometers was phased out by World War II. This tin Mobil Freezone-Mobile Permazone thermometer dates to the 1950s. $450

This En-Ar-Co Motor Oil thermometer dates from the 1950s and features their famous logo. $350

Andrews Oil Company was a local distributor in the Green Bay Wisconsin area. $125

H.K. Stahl were the producers of this Trophy demonstrator thermometer for their motor oils. The two containers at the bottom were filled with different viscosities of oil which enabled the local merchant to demonstrate how they flow at different temperatures. No doubt these thermometers were placed prominently near the oil cabinet at the station. $700

Standard Oil used this large six-foot-tall thermometer in the 1930s. The example shown here is in like-new condition. $850

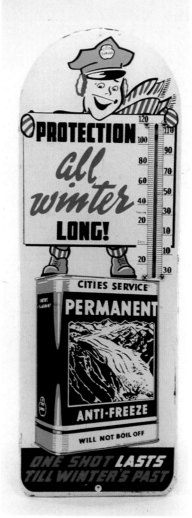

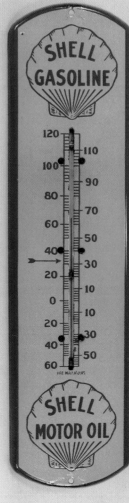

Cities Service were the producers of this Permanent Anti-freeze advertising thermometer. It is made of tin and pictures a can of their anti-freeze showing a mountain scene and glacier. $350

This Shell Gasoline porcelain thermometer dates from the 1930s and features their well-known orange and red color scheme. $850

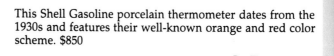

The Texas Company used this square sidewalk stand with a porcelain sign in the 1930s era. The *"Clean, Clear, Golden"* slogan and the accompanying pouring container of motor oil can be found on many Texaco signs dating from that era. The sign itself is not particularly scarce, however, it is difficult to find these in their original stand and base. $600

This die-cut Shell Motor Oil sidewalk sign dates from the 1930s, and is difficult to find in its original stand. $1,000

Union 76 produced this Royal Triton Motor Oil porcelain sign in the 1950s. $250

The period around 1950 saw the discontinuance of lolli-pop style signs to the more favored horseshoe shape. This Golden Shell Motor Oil porcelain sign dates from that era and features their well-known red and yellow color scheme. $750

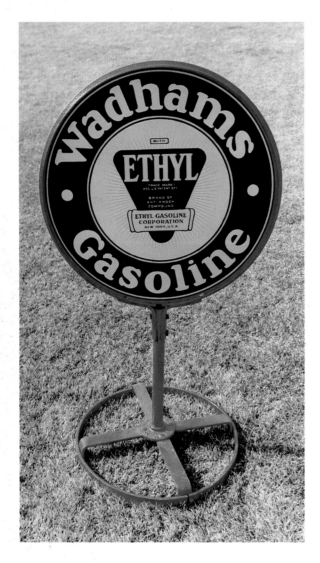

Wadhams Gasoline was a regional company serving Wisconsin, Illinois, Michigan, and Minnesota. The porcelain lolli-pop type sign featured here made sure you knew their gasoline contained Ethyl Anti-Knock Compound. $1,000

White Eagle Balanced Gasoline was advertised on this circa 1930 lolli-pop sign. Notice the eight-sided effect created by the black border around the sign's perimeter. $750

This dynamite duo from Cities Service depicts two of the nation's best-known trademarks of the 1930s-1940s. Both have embossed bases and cloverleaf-shaped frames for their porcelain signs. $750 each

The National Refining Company produced this fantastic En-Ar-Co die-cut street sign in the 1920s. The reverse side is identical. No doubt Royal Gasoline was an alternative choice that the local dealer offered. $1,200

Most all service stations prominently placed the current price per gallon for the motoring public to see. This Polly Gas sign is painted on a tin base which is housed in a wooden frame and is a rare bird indeed. $1,200

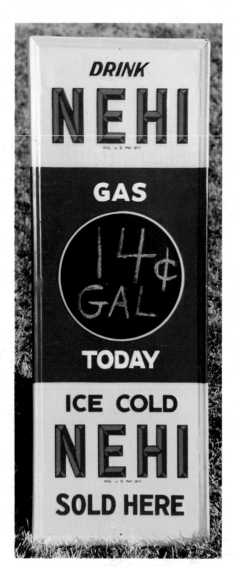

The two pricing signs shown here date from the era around the 1930s. Both are made of tin and advertise what many gasoline dealers considered to be an important sideline - soft drinks. $700 and $2,000 respectively

Winona Oil Company produced this Ivaline Motor Oil tin sign in the 1920s. $75

Veedol also produced tractor oils. This oval sign is made of tin and dates to around 1950. $75

The Midwest Oil Company were the producers of Wil-Flo Motor Oil which is featured on this desirable 1920s tin sign. $1,200

This small-sized 6" porcelain sign was used on company trucks and cars, being fastened to their doors. $1,000

Alemite Motor Oil used this die-cut porcelain sign on a grease gun rack sometime around 1930. $350

This Napoleon Motor Oil sign is made of tin and dates from around 1925. The Van Tilburg Oil Company was one of several producers in the Minneapolis-St. Paul area. $300

A metal banded gas pump globe did the job on this eye-catching Magic Gasoline tin sign dating from the 1920s. $400

Mileage Motor Oil and Greases is featured on this embossed tin sign from the 1920s. $150

The Midwest Oil Company manufactured Ace High Petroleum products. This outstanding tin sign dates from around 1930 and features their race car and monoplane logo. $1,200

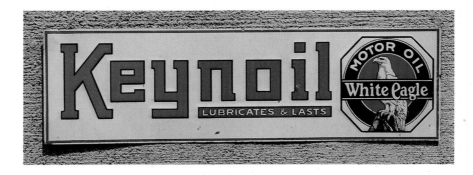

Keynoil produced this White Eagle Motor Oil embossed tin strip sign around 1930. $350

Frazer Axle Grease went the extra mile with this embossed tin advertisement. $650

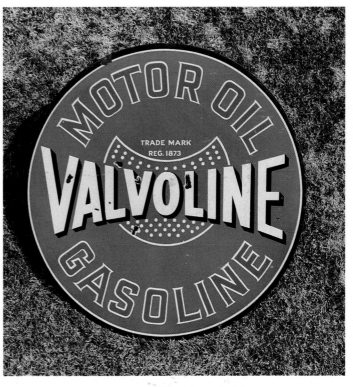

Valvoline's well-known steam valve trademark is featured on this porcelain sign from the 1930s. It measures 42" in diameter and was designed for a lolli-pop sidewalk stand. $750

A closer inspection of the Frazer Axle Grease sign reveals this outstanding picture of early roadside problems. Notice the smiling faces on the trees.

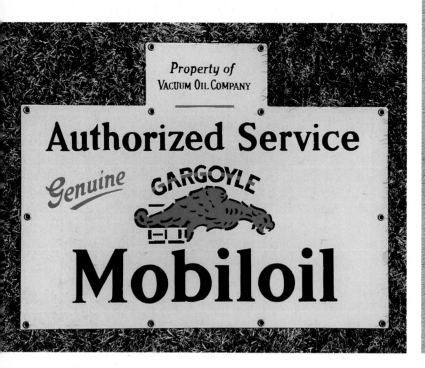

This Authorized Mobil Oil sign was from an oil rack. Each rack would have two signs identical to the one here. A slight variation from other oil racks is the word *"Genuine"* at left. $275

Great graphics are a part of the Conoco Gasoline Minuteman trademark. This beautiful lithographed tin sign dates from the 1920s. $650

United Motors produced this die-cut porcelain sign in the 1930s. Several styles have been found. This one features New Departure bearings on its lower marquee. $1,500

Marathon's famous runner is featured on this porcelain 42" station sign of the 1930s. $1,200

It wasn't all that unusual in the era before 1950 to see service stations selling coal. Here we find an advertisement for just such a station. This Super Quality Shell Coke sign is made of heavy cardboard. It appears to date from the 1930s. Notice that E.J. Wallace Coal Company from St. Louis, Missouri, is listed at the bottom. $85

Several variations of this Weed Chains sign have been noted. The one featured here has a closeup of an automobile's rear tire encased with Weed Chains. The sign also doubles as a pricing box for pre-computer pump calculations by the customer. $850

This tall Wadhams Motor Oil sign is made of tin with a wood frame. It was painted with a reflectorizing enamel. $750

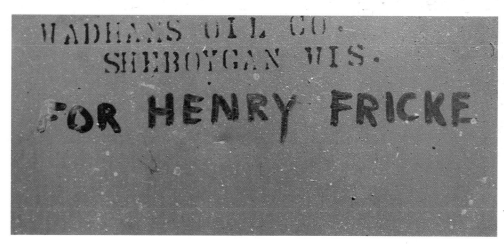

Here's a closeup of the shipping info on the back of the Wadhams sign. $750

As much as this desirable Oilzum sign appears to be porcelain enamel, it is actually tin. $3,500

Several versions of the Phillips 66 30" round shield sign exist. The one here features a large early-style Ethyl burst superimposed over the familiar Phillips logo. This dates to the 1930s. $600

In the 1930s Mobil Oil bought out White Star Gasoline. The large-sized die-cut shield sign shown here was used as their roadside attention getter. $700

Petroleum companies had customized building plaques for their offices. The one featured here is from Standard Oil and originated in Chicago. The number 205 at top could possibly be the street address. $500

Husky Motor Oils used this die-cut steel sign as a highway advertisement to be mounted on poles. It is double-sided, and dates from the 1930s. $1,200

Red Hat Gasoline and Motor Oil were featured on this double-sided 32" porcelain sign from the 1920s. $1,500

Tokheim was one of the world's largest gasoline pump manufacturers. Like any large company, they took care of their customers by offering service on what they sold. The porcelain sign shown here probably dates from the 1950s. $400

Union 76 produced this No Smoking porcelain sign for use at the gas pump island. It dates from around 1960. $350

Supposedly a stash of about twelve of these Pankey Oils signs were found in the 1970s. It features a view of an early bulk truck with wooden spoke wheels. Notice the date 1927 in small print behind the truck's cab. Again, being made of tin, the condition is outstanding considering its age. $1,000

Here is a front and back view of an unusual Cities Service sign dating from the late 1940s. It is made of tin and shows a generic Cities Service advertisement in the photo on top. At the bottom Cities Service gets more specific by telling you that Koolmotor was the way to go. $600

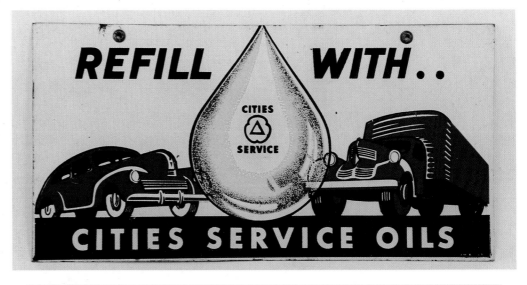

The addition of the Bartles Station attendant at left took what would have otherwise been a rather plain-looking tin sign into a whole new level. Any time you find graphics like these, you're going to have a winner. $900

Even as early as 1950 credit cards were a staple in the consumer market. This D-X sign dates from that era and features their familiar diamond logo. $125

The exact use of this Mobilgas shield sign is uncertain. Most people seem to think it was used on delivery trucks. Possibly an old photograph could reveal its true use. $450

Texaco was the producer of Havoline Motor Oil. This flanged die-cut porcelain sign in the shape of a can dates from the 1930s. $850

Gulf Refining Company produced this colorful flanged sign in the 1930s. Many of these had a tendency to fade over the years. This one is definitely an exception. $450

The era around 1930 saw this 30" porcelain Globe Gasoline sign in service. It was designed to be used in a lolli-pop sidewalk stand. $1,200

Here again we find the colorful Conoco Minuteman on this two-sided motor oil sign. The detail of the minuteman was done by the lithographic transfer process. Needless to say, his presence turns an average porcelain sign into a screamer. $1,200

It's hard to imagine how Globe Refining Company might have made the image area of the can any larger on this flanged two-sided tin sign. The can's design along with their White Seal logo gives the sign's date away as a product of the 1920s. $950

Northwestern Oil Company produced this graphic Noco Motor Oil tin sign in the 1920s. The man at left was an aeroplane pilot, the one at right a race car driver. At the bottom appears the word *"Warehouses"* with the super small print giving their warehouse locations as Superior, Wisconsin; Rhinelander, Wisconsin; St. Cloud, Minnesota; St. Paul, Minnesota; Port Arthur, Ontario; Bemedji, Minnesota; and Duluth, Minnesota. $1,200

R. Swanson was the local distributor behind this Texaco keyhole style sign. It was designed to be mounted either on or near the front door of his Texaco Station. Notice the sign's manufacturing date of 10-3-56 at bottom right. $275

It was a popular practice to feature an oil well derrick as part of the theme on early petroleum signs. This Pennfield Motor Oils sign fits the bill. $175

This unusual Mobil Specialties sign dates to around 1950. No doubt it was part of a larger advertising display. $750

Gulf Oil deviated from their normal orange, white, and blue color scheme to create this Authorized Dealer strip sign dating from the 1930s. $250

Valvoline Motor Oils is featured on this circa 1915 flanged sign. It is unusual because of its departure from their normal colors. $1,000

These two fantastic porcelain signs were used on lubesters and both date from before 1930. The more unusual of the two is the one featured at right promoting Mona Motor Oil. $600 and $750 respectively.

Gold Medal Auto Oil is on the job in this tin self-framed sign dating from the 1920s. Again, the addition of a graphic logo makes all the difference to collectors. $750

Cities Service introduced Koolmotor Bronze Gasolene around 1935. Part of this promotion was this canvas outside banner. The great graphics showing the automobile of that era plus the clock face pump complete with Koolmotor globe put plenty of charisma behind this advertisement. A closer examination of the station attendant's face makes one wonder if Lyndon B. Johnson was a station attendant before he entered the political arena. $750

The cardboard advertisement for Artic-Water shown here makes the statement *"There is nothing like Water for Your Radiator"*. It is interesting to note that the product being advertised is "pure" water, which has been chemically treated. They also boast that their product is safe at 50 degrees below zero. It must have been good. Aside from all this advertising pomp, the great graphics featuring the 1920s sedan along with the artic scene in the background make for a wonderful display item. Note the polar bear walking in the distance. $225

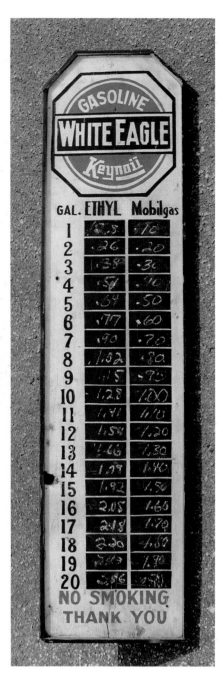

This dual advertisement from the mid-1950s is another example of the cooperation between petroleum companies and after-market manufacturers. Again, like many other advertisements, today's collectors find the background scenery as much of interest as the product advertisement itself. $175

White Eagle Gasoline had this wooden price display placed at an unknown station in the late 1930s. Mobilgas bought out White Eagle around that time and is shown here being priced along with Ethyl. It is of interest to note that White Eagle used many wooden signs through the years which were relatively inexpensive to have manufactured. In contrast to this, their beautiful full-bodied eagle globes were probably the most expensive globes ever manufactured. Imagine the cost of tooling up to build a mold for a White Eagle globe. $1,200

Barnsdall created this huge ten-foot-wide station sign in the 1920s. Aside from the fact it is a little large, it still is a beautiful tribute to one of the world's largest independent petroleum companies who were also the world's first refiners. $750

This large porcelain sign from Husky dates to around 1940. $1,000

Johnson Gasolene is featured on this circa 1930, 30" diameter sign. It was designed to be used in a sidewalk lollipop frame. Shown here is the version with the large early Ethyl burst which was a popular addition to advertising of this era. The other Johnson Gasolene logo features an hourglass in place of the Ethyl logo. $1,200

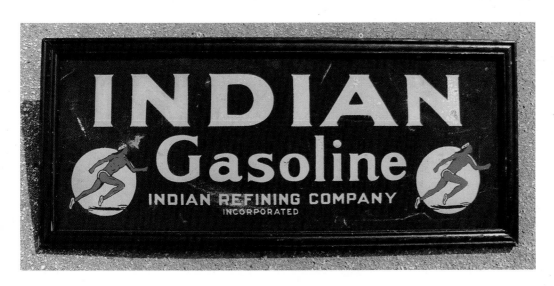

Indian Gasoline is promoted on this very rare sand-painted tin sign with wooden frame. Their logo featuring the running brave was used as Indian Gasoline's trademark from around the mid-1920s until the early 1930s, at which time they were bought out by Texaco. $1,200

It's hard to imagine any way that Edison could have made this die-cut cardboard stand-up sign more appealing. The racing automobile with six large Edison plugs running at full tilt gives the desired impact. This is the kind of piece that collectors stand in line for. $650

Minneapolis-St. Paul had their own local garage association. This tin sign dates from the era around 1950. $275

Few things on earth are as long lasting as the pyramids. Appropriately Pyramid Oils used the slogan *Enduring Lubricants* on this circa 1930s embossed tin sign. $300

This unusual Authorized Station sign from American Oil Company dates to around 1930. It is flat, two-sided, and featured the Amoco logo in the center. $900

Penn-O-Tex was a Minneapolis based company which produced this Rajah Motor Oil sign around 1930. It is made of sheet steel with painted graphics. These advertisements were found away from stations, most notably near city limits, and were mounted with their own individual pole. $1,200

1959 was the year this double-sided tin Invader Oil sign saw service. $400

Few porcelain enamel signs rate as highly to collectors as the Koolmotor Oil beauty shown here. It dates to the 1920s and its unusual shape was designed to go in a matching pedestal for curbside display. $1,800

Plenty of eye-catching graphics are to be found on this Globe's Best Independent Dealer sign. It was designed to be supported at top or for use on an oil cabinet. It is flat, two-sided, and made of tin. $600

Wolverine Porcelain of Detroit, Michigan, was the manufacturer of this beautiful Power-Lube Motor Oil two-sided sign. Many of these were manufactured, but most had a tendency to have acid etching or fading problems. The one shown here is definitely the exception. $950

As the story goes, a local Shell distributor in California had this customized porcelain sign made for his tow trucks in the 1940s. Supposedly only four of these have been found to date. $2,500

Pennzoil wanted to make it clear to the customer that they were getting genuine Pennzoil. To help promote this fact, they produced this two-sided tin sign which listed a $1000 guarantee bond on its reverse side. $175

Here's a look at the Pennzoil Guarantee Bond. Of interest to note in the upper right-hand corner is the serial number 6242 which may indicate that each bond was individually numbered.

Super Shell Gasoline is featured on this porcelain pricer box dating from around 1920. The black area was actually a chalkboard allowing the station attendant the ability to readily change the price. $650

The year 1962 saw this Texaco Credit Card advertisement. The circular sign was made of tin and was designed to be placed in the center of a tire. $175

Skelly Gas Stations were authorized Hood Tire Dealerships in the 1920s. This rare Hood Tire man is approximately 5.5' tall and is made of porcelain. This unusual version differs from most in that it features a completely different man with his left hand raised and his right hand holding the flag. The opposite is true in most of these signs that have been found. Also the man's face and hat are completely different from that of most Hood Tire signs. He also sports a bow tie rather than a conventional tie. $1,800

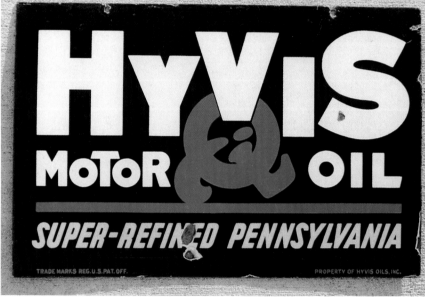

HyVis Motor Oil is featured on this circa 1920s porcelain sign. $225

Rest room signs seem to be coming into popularity as diminishing supplies of other signs make collectors desire what may not have been considered collectible in the past. As can be seen by the photograph, not all rest room signs are created equal. This rare Wood River Men's Room sign dates from the 1930s. $225

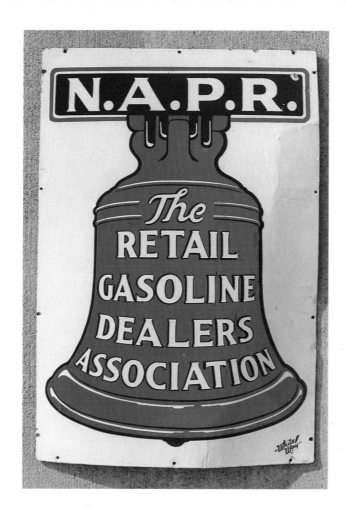

Macmillan Ring-Free Motor Oil tells you what you want to know on this circa 1940s tin sign. $250

The National Association of Petroleum Retailers produced this two-sided sign in the 1920s. It was popular in this era to advertise the fact that you were an independent retailer, thereby conveying the message of good service, quality, and price. $650

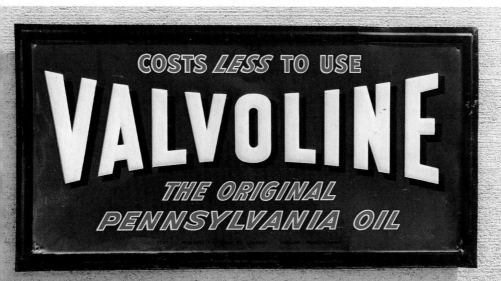

Apparently Valvoline liked to change its color schemes from sign to sign. Here we see a large strip sign made of tin with a metal frame. This time in a paint scheme of orange and black. $850

This embossed tin Valvoline sign deviates from their traditional logo insofar as the lack of its steam valve shown behind the word "Valvoline". $325

Although this Texaco Marine Lubricants sign is of relatively recent vintage, collector demand has kept the price high. It dates from the 1950s and has outstanding graphics. $1,800

Few illustrations could convey a message better than those featured on this Perfect Circle Piston Rings steel sign from the era around 1930. This was one of a set of signs featuring the hog with the bottomless appetite for oil. $1,000

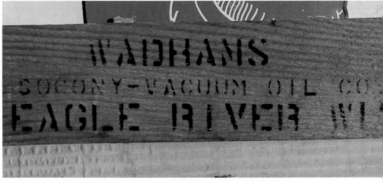

These two photographs show one kind of collector's dilemma. The large Mobil Marine sign with Pegasus logo has been found in a warehouse still enclosed in its original crate and as much as it would be great to uncrate the sign, there is a desire to preserve it as found. Sometimes the true beauty of a collectible is in the fact that you were there first. The photo at the top shows the entire crate, complete with contents. Below is a close-up of the ink stamping on the crate. Because of the Wadhams name, this sign probably dates from shortly after the period when Socony Vacuum bought out Wadhams. We'll take a guess at around 1940. $1,000

This large Texaco keyhole sign dates from 1933. It features their famous black T logo. $500

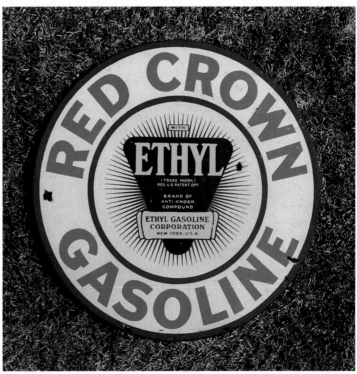

This small Be Square porcelain sign was designed to be fastened to a radiator front. It could be considered an early form of license plate attachment and was no doubt a giveaway at B Square Stations. $200

The two Red Crown Gasoline signs shown here appear to be identical aside from their center image area. The one above shows their famous Red Crown logo and, below, the old-style Ethyl burst which most petroleum companies featured in their advertising. A closer examination will also reveal differences in the letter size which indicates a different stencil used at the factory and possibly even a different manufacturer. They both measure 30" in diameter and were designed to go into a sidewalk lollipop stand. $425 and $475 respectively

How Tydol got away with its orange and black Ethyl color scheme remains a mystery. The Ethyl logo almost always is uniform in appearance, including its colors. Possibly the sign manufacturer did this in error. $375

The yellow and black Ethyl burst adds some pizzazz to this Purol 30" diameter sign from the 1930s. $400

Tide Water Oil was responsible for the production of this Flying A die-cut sign. Its exact use is uncertain but may have possibly been mounted to the sides of bulk trucks. $1,200

Kanotex with Ethyl is featured on this 1930s 30" round porcelain sign. Like most other 30" diameter signs, it was designed to be used at curbside in a lolli-pop type stand. $400

ANYTHING & EVERYTHING

There were literally millions of miscellaneous items manufactured for service station use through the years. Some were meant to be part of service station operations, such as the hats shown beginning on this page. Others were to be sold to the public, like children's toys and puzzles. Still others were giveaways, and these were always a part of the retail marketing of petroleum products.

The following pages will present a brief overview of the many items available to today's collector.

Even as early as the 1920s, kids were attracted to gasoline station giveaways. This Cities Service beanie-style hat was sure to please. $85

Most service stations of past days offered at least minimal mechanical service. The three hats pictured here were worn by gas station mechanics who were coined *"Grease Monkeys"*. Some stations had pits which mechanics worked from. Others had hydraulic lifts. Still others had an area on the side of the garage which was open air. Regardless of where they worked, automobile mechanics often found themselves in a grimy environment requiring the use of minimal head protection. Many petroleum companies offered the local merchant hats to be worn during mechanical repairs which also advertised that station's brand of gasoline products. The Conoco hat is made of leather, the Nicollet Oil Company hat is cloth, and the Mobiloil Special is of a treated cloth. $40, $75 and $50, respectively

It was not uncommon in the era between 1920 and 1960 to find service station attendants in full uniform. This included the matching hats which are shown in the following photographs. The two Phillips 66 hats are each slightly different. The one, on the left, with an open air brow band which was a big help in hot weather. Note the Skelly hat has an additional pen holder above the visor. Values, in rows, top to bottom, left to right. Row 1: $125, $125, $150; row 2: $175 each; row 3: $125, $125, $150; row 4: $125, $125, $175; row 5: $125, $175

There were literally hundreds of petroleum station insignias and logos found on cloth patches made through the years. The following photos give some idea as to the diversity of design. Many of these patches wound up on attendant's uniforms. However, some were also found on hats. All are a reminder of days gone by. Prices range from $10 to $35

Lee was the manufacturer of this 1945 dated Phillips 66 summer shirt. The matching pen was added at a later date. $150

Unitog manufactured this Skelly long-sleeved model. $100

Many times these checkered flags were given away to winners at company sponsored races throughout the country. $175

Buddy Lee produced this silent salesman in the 1930s. These are quite scarce and much sought after by collectors. $350

These small cardboard and celluloid plastic tags were placed on the throttle and choke knobs of automobiles. The local merchant would write down when the oil was changed last as a handy reminder to the motorist. From $5 to $25

By the end of the 1940s most vehicles did not have throttle or choke knobs. These new design lubrication reminders were placed in car door jambs or occasionally under the hood. Most all were made of cardboard or thick paper and had a self-adhesive backing. The one pictured on the far right from Valvoline is actually made of metal. From $2 to $20

This advertisement for Skelly doubles as a fan. Notice that when it is put in the open position, the sedan has proceeded with a green light and the words *"Go With Skelly"* appear in the window. This dates from the 1920s. $125

Dualube and Top Flite Motor Oils were featured on this 1930s ink blotter. $15

Thousands of service stations across the country gave away calendars to promote their products. These three from Shell Oil are typical. $30, $85 and $45, respectively

This stand-up cardboard Shell Tour With Confidence poster no doubt was placed near maps. It dates from the era just after World War II. $125

The 1940 era saw this deck of cards as a promotion of Fleet-Wing Gasoline. $40

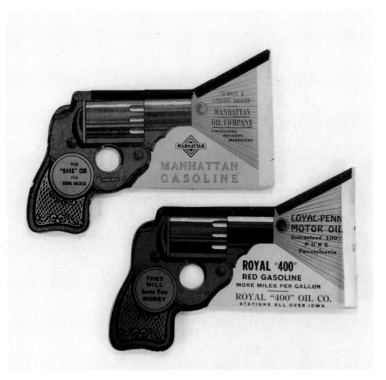

These two pop guns date from the 1920s. $85 each

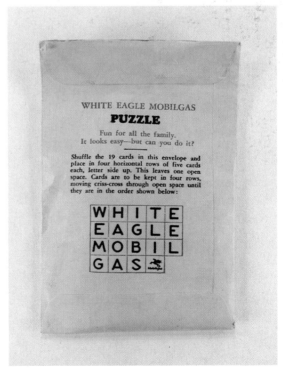

Long trips can be an endurance test on even the most patient parent. This White Eagle Mobil Gas puzzle helped pass the time between stops. $40

Even with three missing pieces, this outstanding Shell puzzle scores big. A few of the major petroleum companies actually sponsored their own aircraft and would go from town to town promoting new station openings with their one-man air show. $150

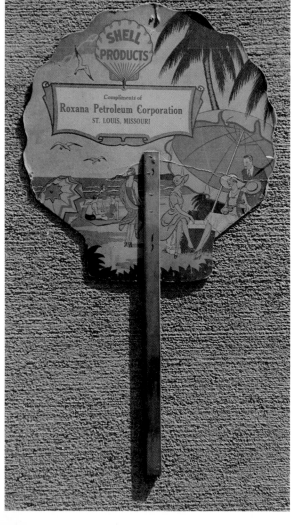

Roxanna Petroleum Corporation makes their statement on this die-cut Shell Products fan of the 1920s. $65

Shell Oil produced this Stop and Go game in the 1940s. $45

Many early advertising giveaways and premiums were designed for children. This pull-out Shell cardboard piece was simple, but got the message across. $40

Plenty of the giveaway market was targeted at household type items. This Superior Needle Book from Skelly contained 60 needles and a threader. $15

Believe it or not, there was a time when maps were actually given away at service stations. This touring map rack had plenty of room to help you find your way no matter where you were going.
$225

The lack of freeways up until the 1960s put the pressure on local gasoline merchants to offer travel information. This service was welcomed by the traveling public and normally resulted in a gasoline or motor oil purchase. $65

Few other long-term promotions were as successful as the petroleum companies' free map distribution program. The following photographs show just a few of the many graphics that can be found on maps. The beautiful White Eagle 1932 road map, above, depicts their famous globe atop a visible pump. A closer look can be found in Chapter 1: $75. At left (l-r): $10, $15, $20, $20; below: $15, $20, $18, $20

92

Outstanding graphics abound in this 1920s road map of Minnesota from Fyre-Drop Gasoline. $75

Here's the Fyre-Drop map opened up. As a service to the public, many petroleum companies made it a point to index the location of their service stations throughout the state. The red dots represented here give you an idea as to the extent of Fyre-Drop Gasoline in Minnesota.

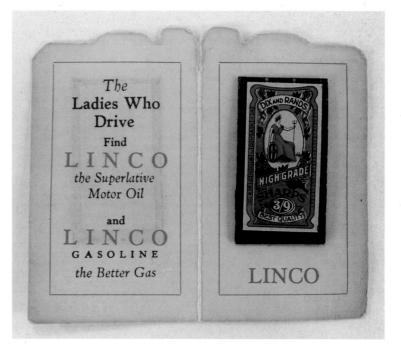

This small Linco needle kit took on the shape of their famous motor oil can. Note that small as it is, it no doubt was well received as a giveaway. $40

The year 1937 saw this Major Efforts indexer from Texaco. This was given away to gasoline retailers in an effort to help pave the way to better sales. $45

Service station owners were constantly being bombarded by the petroleum companies with promotional ideas. This outstanding example depicts Santa visiting his local Skelly dealer on a dealer's sample Christmas card. $20

The graphics on the box which contained this tin Sinclair Gasoline match holder pretty much size up the use and years that this was given away. $300

Gold Medal Oils is featured on these two circa 1920s ink blotters from the Kunz Oil Company of Minneapolis. Their logo with the slogan *Best For The North* can be seen on the right side. $15 each

Shell Oil produced this household kit to be sold at service stations. Notice that Shell Oil took every available opportunity to include their logo inside and out. $225

Royal 400 Gasoline makes a big impact on this circa 1920s cardboard sign. Although its exact use is uncertain, the four grommets on the logo's perimeter indicate its possible use as a winter front for a truck. Regardless, it scores big with collectors. $400

Unlike automobiles of today, the vehicles of the 1920s had problems maintaining high temperatures in cold weather. These waxed cardboard radiator covers were just the ticket to maintain good operating temperatures in northern latitudes. Most all the major service companies offered these as a giveaway which collectors of today call "winter fronts". The Cities Service one without the black band has an original merchant's guide page which is shown advertising two of the covers. The fine print below reads: *"Radiator Protectors. When the chill winds start to blow and the thermometer nosedives, motorists are always pleased to get the added protection furnished by a radiator shield. These protectors usually stay on a car all winter. Everywhere the motorist drives your product is being advertised."* Pg. 96: $35 each; page 97 (top to bottom): $50, $100, $75

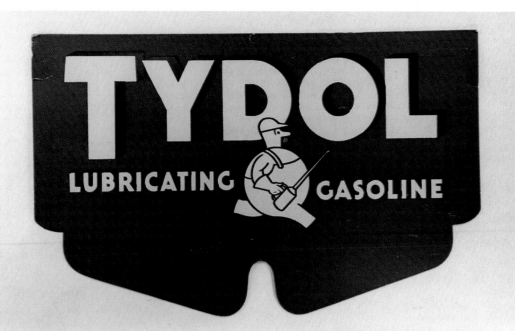

To get the most bang for the buck, filling stations advertised their products on matchbook covers. There were literally thousands of different ones issued through the years, each a silent salesman to the public. Even today, matchbook giveaways can still be found at some service stations. The sampling in the following photos will give you an idea as to the diversity and beautiful graphics found on matchbook covers. Page 98 (top to bottom): $5, $20, $20, $20. Page 99 (clockwise from top left): $15, $15, $12, $5, $15, $8, $8, $8 (center)

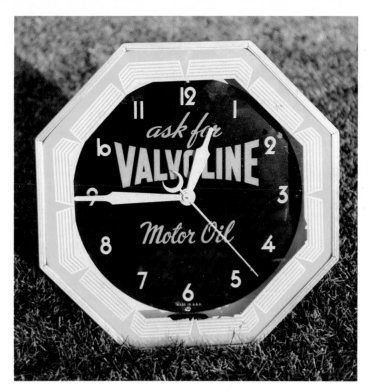

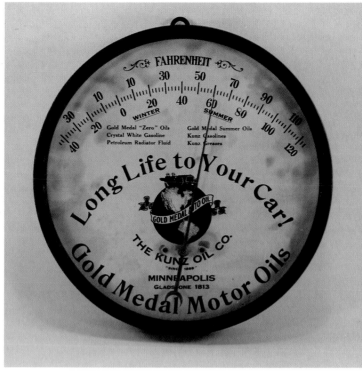

Neon Products of Lima, Ohio, was the manufacturer of this Valvoline eight-sided neon clock in the 1940s. Several versions of this clock have been seen, this one featuring a green dial. Notice remnants of the 1930s era can be found in the deco-style 10, 12, and 2 positions. $600

This station thermometer was in service at Kunz Gasoline Stations around 1915. Considering its age, the paper dial has come through rather well. Notice the globe logo with the early touring car on top. $225

Many early service stations were small buildings that did not require large heating units. This fantastic clamshell-shaped gas heater is from a Shell Petroleum Station in the New England states and dates to around 1920. Both sides are identical and feature multi-colored glass jewels which were inserted through the body. These would flicker when the burner was in operation. The Shell embossing can be seen at the bottom. $1,500

Rex Oil Company of Minneapolis took this wide-angle shot
of their operations in the mid 1930s. Although one of the two
visible pumps in the photograph has a globe, to this day
none have been found. $125

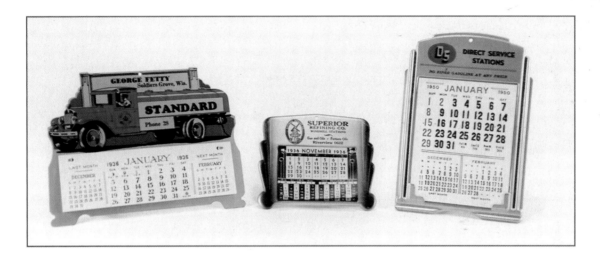

Shown here are three desk calendars. The outstanding
Standard calendar at left features a die-cut truck with
Standard Oil's early logo on its door. $200, $35, $25, respec-
tively

Westland Oils pottery cowboy hat ashtray is pictured at left.
At center is the well-known Jenney Aero Gasoline ashtray
made of tin. At right is an ashtray featured by a local distributor from New Richmond, Wisconsin, that was a retailer
of Cities Service products. $85, $125, $25, respectively

These two Hood Tire giveaways were part of the marketing strategy of the 1920s era. The Hood Man at right is on a suction cup and could be mounted on a car's dashboard. The suction cup: $225; ruler: $50

Wooden matches were kept in these tin match container boxes from the 1930s era. Each has a small suction cup mounted on its back side so they could be dashboard mounted. $85, $65, $150, respectively

These hat badges were worn by employees of distributorships and bulk plant operations. Each has a personalized embossing showing which company it originated from. The Milo at left and the Regal Oil at right also have the added feature of employee numbers. They date from around the 1950s. L-R: $125, $150, $85

Most every major brand retailer of the 1930s era had their employees wear name badges. This foursome is typical of the dozens that were manufactured through the years. Conoco, $250; Texaco, $175; Skelly, $200, Cities Service, $350

These two items are plant badges. The one at left is from Standard Oil Company out of Neodessa, Kansas, and is made of brass. It dates to before 1920. At right is a 1930s Cities Service Refining Corporation badge which could conveniently hold the employee's photograph. $85 and $175

Petroleum companies were constantly on their retailers to better their sales of tires, lubricants, automotive parts, and other products besides gasolines. This brass cigarette holder was awarded in the 1940s era to Mobil dealers for balancing their sales between gasoline and other automotive products. $175

A sizable collection of insecticide sprayers could be assembled by anyone who looks for them. Cities Service used this handy model in the 1940s. Notice that the holding tank is embossed with the Cities Service logo. $125

Here's one more item to be added to that long list of products designed to help the motorist travel. Spinning the knob at right would enable the viewer to see the mileage from his point to the city shown in the screen. This unit dates from the 1940s. $1,000

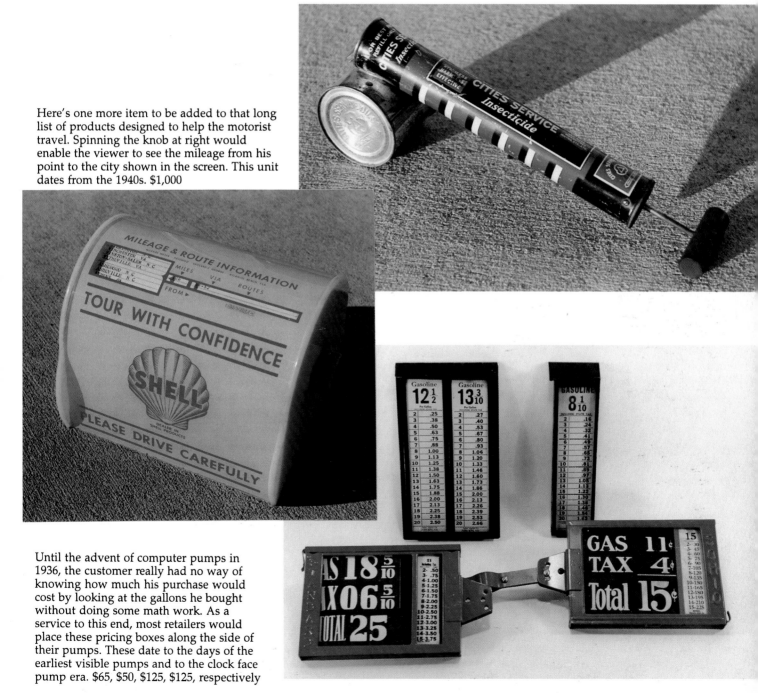

Until the advent of computer pumps in 1936, the customer really had no way of knowing how much his purchase would cost by looking at the gallons he bought without doing some math work. As a service to this end, most retailers would place these pricing boxes along the side of their pumps. These date to the days of the earliest visible pumps and to the clock face pump era. $65, $50, $125, $125, respectively

Suggestions for better restroom facilities could be mailed in with the help of this prepaid card holder from Sinclair. It dates from the 1950s. $125

The 1950s saw the use of this stand-up cardboard Radiator Flush display. "Easy to use" must have been true as the man in the sketch is dressed in a suit. $85

Transporting auto batteries around was made easier with the help of this leather and metal carrying strap dating from the 1940s. $25

Shell Oil had this wood, brass, and silver cigar holder made in the 1940s. $325

This floor sweeping receptacle may not have been a giveaway item at Don Miller's Filling Station. If it wasn't, it probably was offered to the public very inexpensively. Both Mobil and Wadhams can be seen on the advertisement which would date this piece to the late 1930s. $125

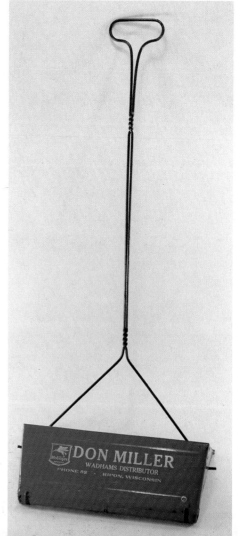

Even restroom keys have become an item that's not overlooked in today's collectibles market. The two from Texaco are made of tin, the rest of plastic. The ones pictured here date to the 1950s. Texaco: $100 pr.; Pure: $150 pr.; Skelly: $50 pr.

Mobil presented this service award to local dealer Arthur Peters in the 1940s. It is made of painted plaster. $75

Paper weights were a popular desk accessory in the 1950s. The National Refining Company offered this solid brass model in 1957. $125

Here's a shot of the reverse side.

The following three photographs show measuring sticks that were used in the early years of motoring. Previous to the advent of automobile fuel level gauges, a measuring stick was required to see exactly how much gasoline was left in the tank. Hundreds of gasoline retailers offered these free of charge. From $25 to $75

These two first aid kits were offered by Shell. The one at left would be found in service at a bulk plant operation. The one at right was used in a tank truck or local retail station. $85 and $55, respectively

This small Flying Red Horse first aid kit was a giveaway item from Mobil. It dates to the era around 1950 and could be conveniently stored in the car's glove compartment. $40

A promotion by Shell earmarked this watch for distribution to Shell dealerships in the year 1940. Shown is the watch's back side which had a clear crystal in which the movement could be seen. Notice the large main gear at right has embossed the word *"Shell"* on its surface. $450

The famous En-Ar-Co boy is pictured here on two giveaways. The stand-up cardboard calendar is from 1939. The small thermometer at right dates from the 1930s and was designed to be mounted on the visor. $110 and $35

Bath time was fun time with this Sinclair Heating Oil miniature truck made of soap. Although it was targeted for the children's market, possibly an occasional adult had tub time fun. $35

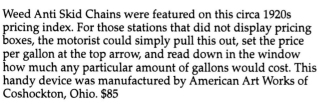

Weed Anti Skid Chains were featured on this circa 1920s pricing index. For those stations that did not display pricing boxes, the motorist could simply pull this out, set the price per gallon at the top arrow, and read down in the window how much any particular amount of gallons would cost. This handy device was manufactured by American Art Works of Coshockton, Ohio. $85

The 1950s saw the production of these two children's items. The one at the top being a bank from Sinclair in the shape of Dino the Dinosaur, while the item below is a plastic toy truck from Esso. $30 and $65, respectively

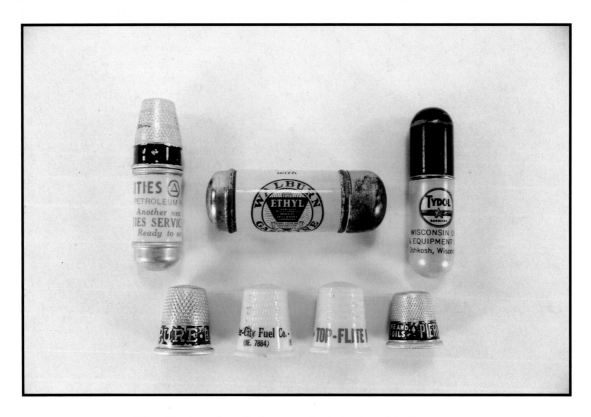

This collection of small giveaways includes four thimbles at the bottom, a sewing kit from Cities Service at left, a sewing kit from Walburn Gasoline with a large early Ethyl burst at the center, and at right is a Tydol Flying A Gasoline cigarette lighter. Top row (l-r): $25, $40, $40; bottom row: $15 each

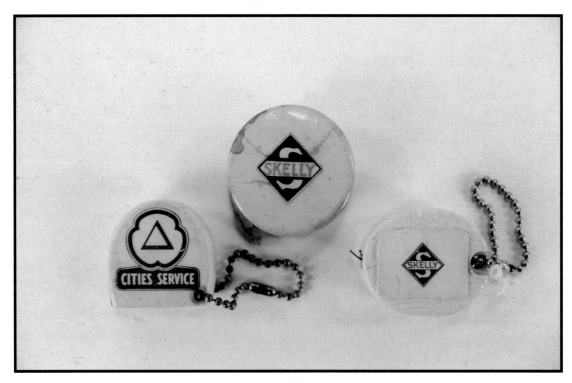

One more item for which gasoline retailers found willing takers was a household tape measure. The Cities Service at left is made of plastic with a metal tape dating from the 1950s. The round Skelly tape measure at center is a celluloid plastic over tin and has a cloth tape. This dates from the 1930s. The Skelly at right is plastic with a metal tape and dates from the 1950s. $35 each

Assembled here is a collage of miscellaneous pin backs. These were given away for a variety of reasons. As an example, the orange Pioneer Service Station pin at top left was designed to be placed on the visor indicating when the oil was changed last or the car was greased last. Skelly had the large red pin designed for their 20th anniversary in 1939. Possibly the oldest pin in the grouping is the Polarine at center right showing Standard Oil Company's original Red Crown logo. Between $35 and $125

Sure to drive mom and dad crazy on those long trips, we find here an assortment of whistles and clickers. All these date from the 1920s, long before car radios offered alternative entertainment. $35 each

These handy drinking cups could be conveniently stored under the seat and pulled out at the next water stop. They date from around 1920. $350, $50, $150, respectively

One of the best looking gasoline station giveaways that might be found is this Socony Motor Gasoline celluloid-back paperweight mirror. It measures approximately 4" in diameter and dates from the 1930s era. $125

Alemite produced this convenient tin die-cut countertop display rack for their grease fittings in the 1920s. $100

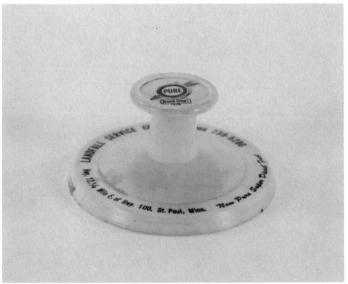

Most people would cover the valve stems on their automobiles with screw-on threaded caps. The 1950s saw the introduction of these valve stem caps in the shape of crowns issued by Standard Oil. Although the automobile market was the intended target of this promotion, surely station attendants were bombarded by kids wanting these for their bikes. The penny has been included in the photograph for size comparison. The top row represents their famous Red Crown brand gasoline, the bottom row signifies Gold Crown. $30/set

Ever see one of these? This is the type of thing you could take to a party and play "Whatzit". It's actually a windshield ice scraper dating from the era around 1960. $40

Lyle Lindquist was Edgar Oil Company's local distributor in Ashland, Nebraska. This note board clip dates from the era of the 1940s and shows a delivery truck of the day hauling Skelly brand petroleum products. $65

Western Fuel and Oil Company used this desktop perpetual calendar in the 1930s. At the bottom, notice the Husky dogs which are engraved at each side. $150

This United Service Motors pocket mirror features their famous logo. It dates to the 1930s. $95

Mobil Oil gave away this suction cup style coat hook in the 1940s. $35

The three items pictured here were used to remind motorists when to change their oil. The one at left is from Sinclair and is visor mounted. The center one is from Franklin Oils and is dashboard mounted. The one at right is from an independent garage that advertised "everything for your car". All three of these have small rotating wheels with the numbers zero through nine etched on them. These could be turned at will so that the mileage at time of lubrication was recorded. $40, $100, $25, respectively

At top left is a Cities Service truck flashlight keychain dating from the 1950s. At center top is a plastic model race car advertising Jet Action Tydol Ethyl. This dates from the 1930s and was designed to have a balloon inflated through its top with exhaust coming out the back of the car. At right is a toy spinner from Gold Medal Auto Oil. The white pencil is from Standard Oil and has its original white crown eraser still attached. The French Auto Oil pencil was manufactured for the Marshall Oil Company of Iowa and dates to the 1920s. Back row (l-r): $150, $100, $20; back pencil, $15; front pencil, $10

As a service to farmers, this Delivery Ticket box from Standard Oil was left in a conspicuous location and could be set to tell the driver what products were needed. Notice the clock-type hands set at pressure grease and heater oil. This unit dates from the 1940s. $150

In 1984, Phillips 66 awarded this plaque to those dealers with high volume oil sales. $125

A full box of household handy oilers dating from the 1930s is pictured here. The containers used what is known as leaded tops and are still complete with contents. $175

Mobilgrease was dispensed in this early container with the help of the device pictured here. It was designed to be pushed through the can for a quick load of FS Mobilgrease. $85

Powell Oil Company of Marshall, Minnesota, produced this small thermometer mirror combination in the 1940s. $75

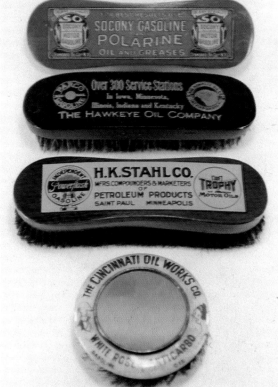

These four thermometers were all giveaways at the local petroleum retailer. The Golden Shell was designed as a visor clip-on thermometer and dates from the 1930s. The other three are household thermometers and date to the 1940s. $100, $85, $125, $45, respectively

The four brushes shown here give the impression of shoe buffers in a polishing kit. However, these are actually brushes dating from the 1920s and 1930s for removing lint from mohair car interiors. $125 each

There were literally thousands of service stations advertised on ashtrays in the 1930s, 40s, 50s, and 60s. The three examples presented here would be typical. $20 each

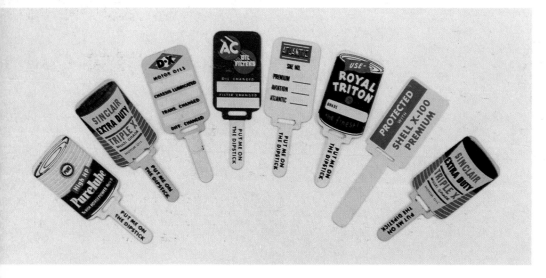

Even the kitchen table couldn't escape the non-stop promotion of petroleum products from the major manufacturers, as shown by these salt and pepper shakers. Left to right per pair: $60, $35, $40

The mileage at each oil change could be conveniently notated on these dipstick tags dating from the 1940s and 50s. $20 each

These desk paper weights promoted, from left to right: Husky, Red Crown Gasoline, and Lion Petroleum. $125 each

The 1930s through 50s saw tens of thousands of cigarette lighters being handed out in company promotions. From $35 to $75 each

Mobil produced this unusual looking Bug-a-boo Moth Crystals container in the 1940s. $135

Children's banks were a popular giveaway through the years. The Mobil Oil baseball bank at left is made of glass. The Phillips 66 *"fat man"* is constructed of plastic. The cans are made like their true-life counterparts with cardboard sides and tin ends. The Sinclair H-C Gasoline bank at right is made of tin.
Top row: Fat man, $85; Tower, $65.
Bottom row (l-r) $50, $40, $35, $35, $65

Few items were handed out as station give-aways during the war years. The shortage of metal and personnel for such promotions made it difficult to take on a large scale campaign. This 1944 calendar from Johnson Oil Products is made of cardboard. $75

Oil changes could be recorded on this metal visor mount mirror from the Leonard Nelson Service Station out of Willmar, Minnesota. $45

These two metal clipboards date from the late 1940s. The one at left features Illinois Oil Products and, at right, we see Brilliant Bronze Stations, part of the Johnson Oil Products Corporation. $65, $45

No doubt pen giveaways were issued in quantities which may have only been surpassed by matchbooks. This popular practice which would show the local station's name and address has left an endless flow of different pens and pencils to be collected. The items that are shown here are typical of the thousands that are available. The most sought after are those which contain samples of oils, have oil cans on top, or are from the more rare and desirable service stations. $15 to $25

If collecting smalls is your thing, you'd be delighted to own this circa 1959 salesman's sample kit from Skelly Oil Company. These were taken with the local Skelly rep to the service stations on his route so that his local merchants could view the latest giveaways, promotions, and point-of-sale items available. $500

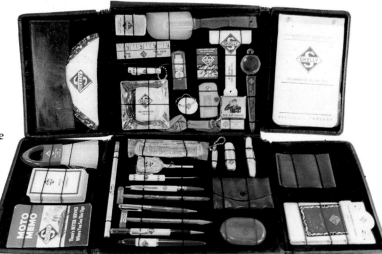

The 1950s era saw the production of tens of thousands of pole-type thermometers. Each of these were made of plastic and were meant to represent the familiar roadside signs of America's service stations. All the major oil companies had these as did many independents. Some were designed for use strictly as thermometers. Others had a calendar attached to their bottom. From $35 to $100 depending on rarity

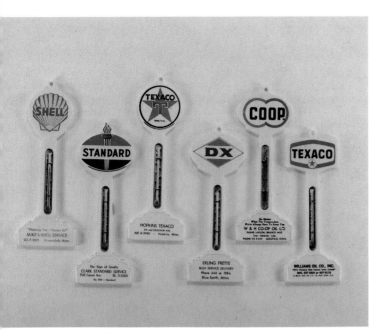

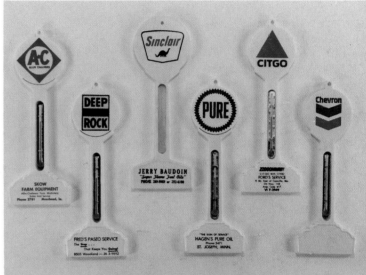

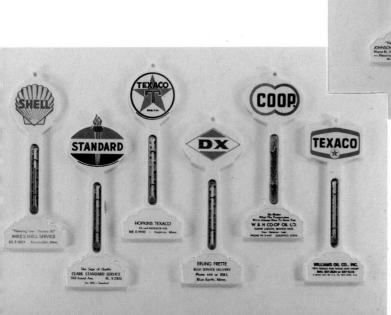

The four photographs included here represent an era which ended around 1950. These license plate attachments were given away at thousands of stations across America. Some were painted, others used a reflectorized material. From $75 to $200

"The greatest achievement in motor lubrication" was presented on this circa 1930s motor oil demonstrator. Each glass tube was filled with a different grade of oil which could be turned upside down providing a clear look at each grade's flowability. $125

These two demonstrators by Shell Oil were designed to have oil poured from the full bottle into the empty bottle. In theory this sounds like a great way to demonstrate, but it was probably a constant clean-up problem in practice. $300 and $325

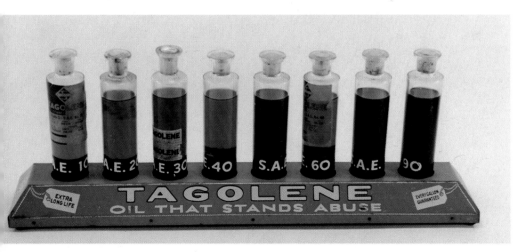

A slightly different demonstrator for motor oil is shown here in this Tagolene kit by Skelly. The bottles were designed to be individually removed for a personal inspection by the customer. $125

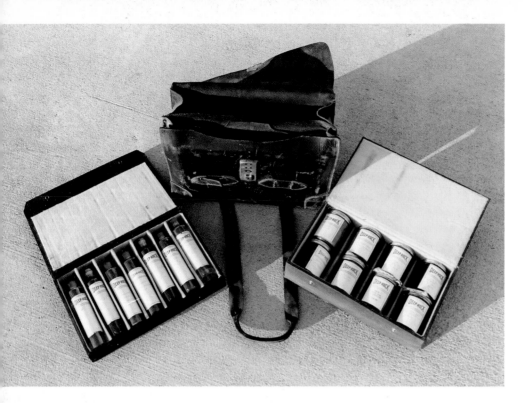

Salesmen carried this demonstrator kit with them on their routes. The containers at left were oil filled; those at right contained grease. This complete original kit dates from the 1930s and is from Deep Rock. $85

The following photographs show some of the personalized company locks that were in service between 1920 and 1950. Some companies put their name in letters on the locks. Other companies put their logo. All of the ones photographed here have brass bodies. These could be found at service stations and on storage buildings and sheds. They were also used at bulk plants on fences and other facilities. From $30 to $100 depending on rarity

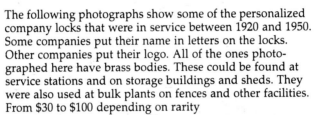

This Shell Oil tanker dates from the 1940s and is painted in their typical red and yellow color scheme. $600

The 1960s saw the introduction of this tractor-trailer Texaco tank truck made of tin. $250

Sinclair H-C Gasoline was the theme on this 1940s tank truck. $400

This beautiful Cities Service tow truck with operating weight
mechanism dates from the 1950s. It has original paint and is
quite scarce. $300

This beautiful 1920s style tank truck hauled generic gasoline,
but the brand really wasn't important to the lucky child that
owned it. $1,500

This well-used 1920s tanker from an unknown company has seen better days. Regardless, it's a welcome addition to any miniature vehicle collection. $400

Wyandotte Toys manufactured this gasoline truck in the 1940s. $400

Shell Fuel Oil is featured on this 1930s tank truck manufactured by Buddy L. $650

CHAPTER FOUR
OIL CONTAINERS, DISPENSERS, AND PUMPS

Second only to gasoline, oil marketing was a breadwinner at every service station. The mechanical nature of the automobile proved to be reason enough for petroleum companies to "see gold" in every vehicle. An entirely separate field of collecting can arise in oil related containers and advertising. Thousands of cans and bottles have surfaced in the collectibles market, along with bulk containers, shipping crates, racks, lubesters, pumps, and the like. The list seems endless.

Along with early tin-type oil cans, many service stations of the 1920s utilized tall one-quart glass oil containers as dispensers. This rare 1920s Crown Cork and Seal Company capping machine is shown with a typical tall one-quart oil bottle from Shell. Pushing down on the handle at right would lower the cap and firmly crimp it to the bottle top. $250

Double Shell SAE 30 Lubricating Oil was dispensed from this unusual can featuring a large Shell logo. $125

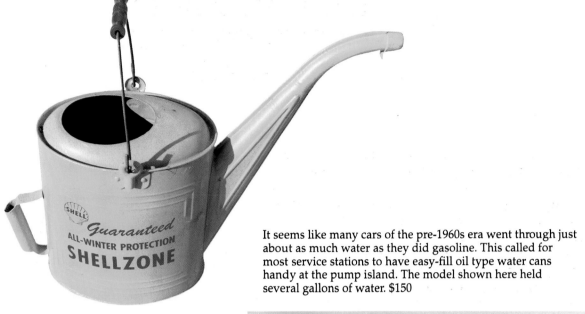

It seems like many cars of the pre-1960s era went through just about as much water as they did gasoline. This called for most service stations to have easy-fill oil type water cans handy at the pump island. The model shown here held several gallons of water. $150

Motor oil got its fair share of counter space with the help of this typical Cities Service display rack dating from the 1930s. $425 complete with cans

Sixteen Tiolene Motor Oil bottles were contained in this rack. Tiolene used a bottle which was taller and more narrow than many other oil companies' bottles, but still contained one quart. $750

Two different types of bottle racks are presented here. The one on the left was more designed for stationary storage as it would no doubt be fairly heavy when full. The one pictured at right was quite portable, offering the station attendant the ability to easily transport it to pump-side use. $1,000 and $750, respectively

In later years, the tall one-quart bottles gave way to more conventional shaped containers. This may have been due to the need for a built-in funnel system to facilitate pouring. The rack pictured in these photos contains three one-quart bottles with screw-on funnel tops. The photo at left shows the complete rack with a rare small-sized Mobil Oil shield sign attached. At right you can see how convenient it would be to lift out up to six quarts of oil and transport it to the vehicle. $350

This rare Shell Motor Oil stand was used as an outdoor display rack for tall-style one-quart oil bottles. Notice a duplicate porcelain sign is mounted to the back side. $1,800 without bottles

Curbside display was an important marketing point for motor oil. This 18-quart display rack dates from the era around 1930. $700 complete with cans

Other styles of oil dispensing were tried in one form or another through the years. These early Shell Oil Company dispensers are made of solid copper and could contain between one and six quarts of oil. All date to before 1925. Left to right: $250, $150, $150, $250

Depicted here are six typical examples of one-quart oil bottles all having screw-on funnel-top pouring spouts. The aqua Standard Oil of Indiana pictured second from the left is quite rare and contained Polarene 10W Motor Oil. From $40 to $85

Five gallons of Wadhams Motor Oil could be contained in this bulk can made by Alisco of Philadelphia in the 1920s. Many farmers and other bulk users found these containers a convenient way to get oil to their operations. $50

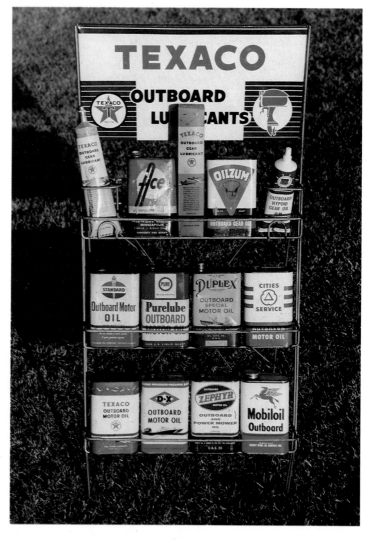

Targeting specific markets such as boaters could mean additional big business for retailers. The Texaco Outboard Lubricants rack presented here contains different manufacturers' products that all relate to some type of marine lubrication. $750 complete with products

Phillips 66 utilized a porcelain enamel sign on this motor oil rack dating from the 1940s. $750

Colorful graphics are found on this wooden motor oil box of the 1920s. Texaco's image during trans-shipment must been important as this box was not normally sold as a retail container. $1,000

Roxanna Petroleum Corporation used this wooden box to ship grease to its retailers. $75

This portable wire-frame rack was used by Mobil Oil to dispense their Lubrite products in the 1940s era. $425

Invader Motor Oil used this tiered display rack in the era around 1930. The porcelain enamel base shows its colorful logo on the left with large Invader letters below. Of interest to note is that the local distributor was fired right onto the porcelain, making a permanent record of its origin for collectors. $900

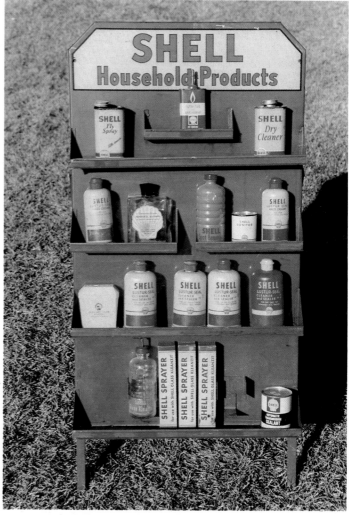

Even the household market couldn't escape being targeted on this 1930s Shell products rack. $350

Kerosene lamps as well as automobile headlamps burned kerosene. This Perfection Oil dispenser contained kerosene fuel and dates from the era around the turn of the century. $650

As with many utilitarian objects from early station use, the Shell Household Products rack came complete with this notice prominently placed on its back side.

Five miscellaneous one-quart oil bottles are shown here. The three to the right have fired-on graphics. $50, $60, $125, $45, $50, respectively

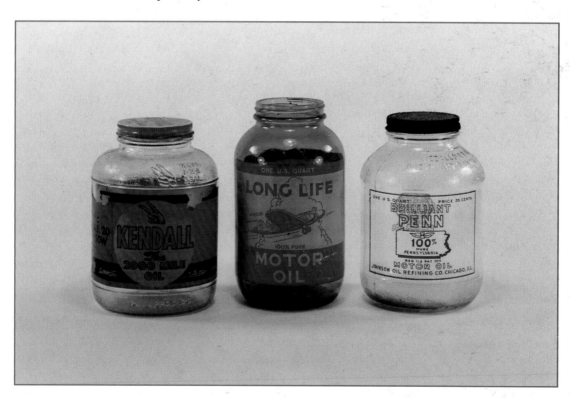

Several manufacturers produced glass mason jar style oil bottles during the war years because of the shortage of metal. $35, $50, $50

This pyramid-topped lubricating oil can dates from the 1920s. $175

Outstanding graphics make this one-gallon French Auto Oil can a winner. The racing scene on the front side is contrasted by the discussion on the reverse of four various grades of oil. $1,200

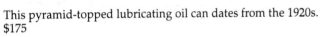

Most anyone would want these graphics in their can collection. As utilitarian as they are, oil cans can be a work of art.

This grouping of one-gallon oil cans all date from the 1920s era. The terrific graphics found on the Sinclair Opaline Motor Oil can are enough to make anyone start collecting. The author wonders if a porcelain enamel sign such as is mounted on the radiator of the race car exists. Top row (l-r): $100, $325, $400; bottom row: $1,000, $200, $400

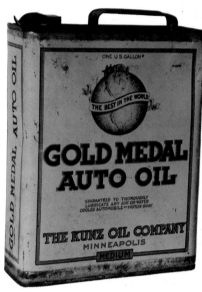

Drum containers offer their own unique appeal in the collectibles market. Like most other petroleum-related collectibles, these five-gallon containers vary in graphics from company to company. $85 - $150

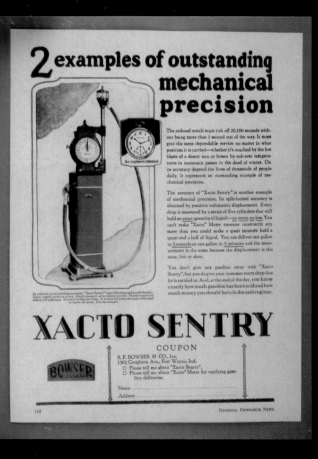

The earliest oil cans were manufactured with leaded soldered seams. The three shown here date from the era around 1915 to 1920. The New Life can at center contains automobile polishing compound rather than oil. $125, $200, $225, respectively

At first glance, this Primrose Speedway Motor Oil container may appear to be a one-quart container. A closer examination will reveal that it holds five quarts of motor oil. $275

White Eagle Refining Company produced the Keynoil Motor Oil shown at left. The early Pure Oil Company can at center dates from the teens. Five pounds of grease were inside the White Eagle container at right. L-R: $275, $150, $75

The Pep Boys have been around since the 1930s. Although they did not operate as a petroleum company, they did distribute motor oil and hundreds of other after-market items in their retail stores. Their products represent an American success story considering they still are in operation today. $250

Anti-freeze was also marketed in one-quart containers for many years. L-R: $40, $50, $65, $65, $85

The following grouping of early half-gallon oil cans presents itself with the usual colorful graphics. Most notable is the Trop-Artic from Manhattan Oil Company with its beautiful artwork. The yellow can issued by H.K. Stahl Company uses the word *"Blitzen,"* which is German for lightning. Top to bottom, left to right: pg. 140: $175, $175, $150, $750, $750, $400; pg. 141: $300, $250, $225, $175, $325, $325.

There were literally thousands of oil cans manufactured through the years. The following photographs depict your not-so-average looking cans. Each were selected for inclusion here based on scarcity and condition. Can collecting has become one of the fastest growing segments of the petro collectibles market. The top two groups on page 143 are oil from Minneapolis-St. Paul distributors.

Values, top to bottom, left to right by rows.

Page 142, row 1: $150, $45, $150, $85, $150; row 2: $200, $12 $225, $85, $450; row 3: $125, $300, $85, $125, $65; row 4: $225, $125, $225, $175, $125.

Page 143: Row 1 top: $100, $475, $550, $125; Row 1 bottom: $150, $100, $175, $125,. $75; row 2 top: $135, $650, $600, $750; row 2 bottom: $450, $500, $850, $850, $750; row 3: $125, $250, $225, $175, $85; Row 4: $250, $175, $350, $250, $225

page 150: row 1: $500, $225, $300, $250, $250; row 2: $500, $250, $250, $175, $125; row 3: $125, $225, $150, $500, $150; row 4: $150, $125, $125, $135, $225

144

The following photos show eight five-quart oil cans. Most all these date from the 1930s era and considering their age have been preserved quite well. Left to right: Top row: $150, $65, $350, $375; bottom row: $350, $350, $175, $125

Egyptian pyramids were the logical choice for this Pyramid Oil Company grease can of the 1920s. Several other companies manufactured these ten-pound pail-style cans, each with their own logo. $85

Cody, Wyoming, was the home of Husky Refining Company, which produced this ten-pound can with bale in the late 1950s. $225

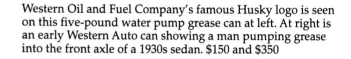

Few logos in the petroleum collectibles market are as sought after as the Conoco Minuteman featured on this five-pound Auto Grease container. $400

Western Oil and Fuel Company's famous Husky logo is seen on this five-pound water pump grease can at left. At right is an early Western Auto can showing a man pumping grease into the front axle of a 1930s sedan. $150 and $350

This closeup shows the detailing and overall artistry involved on the Minuteman logo. It is no wonder he is so popular.

Anybody who has worked with mechanical things knows the necessity of good lubrication. These one-pound grease cans were a handy addition to the garage or farm. Notice the top row has lids which covered the outside of the can. The bottom row uses push-in type lids. L-R: Top $75, $160, $135, $160; Bottom: $65, $225, $100, $40, $100

Containers were needed for most everything relating to the automobile. Here's a grouping of five miscellaneous containers. TL: $35; BL: $40; C: $85; TR: $35; BR: $40

The different containers used for lubricants seems to just go on and on. TL: $125; BL: $50; C: $35; R: $75

Hood produced this tire patching rubber cement called Suprex in the 1930s. It offers views of the famous Hood Tire Man on its side and front. To the right is a Gold Medal Grease can from Kunz Oil Company of Minneapolis. $200 and $150

At left is a Cities Service container of lubricant designed for sticky doors. The two containers right of that show a can of radiator stop leak from the 1920s on top of a small Johnson Oil Grease can. The waxed cone-shaped container advertising Clark Super Gas was a promotional giveaway and was filled with sand. These were handed out during the winter months and were designed to be stored in the trunk of your car. Johnson's Prepared Wax not only cleaned your automobile but also fine furniture, woodwork, and linoleum floors. What a concept! Below that are the Pep Boys featured on this Pure As Gold one-pound container of cup grease. L: $20; CL top: $40; CL bottom: $75; cone: $85; TR: $85; BR: $100

Sales of household oils could add up to big dollars for the local retailers. These two photographs will give you an idea as to a few of the styles available through the years which have become known as handy oilers. Top row (l-r): $85, $85, $75, $25, $30, $20, $25; bottom row: $25, $25, $25, $20, $25, $25, $25, $25, $25

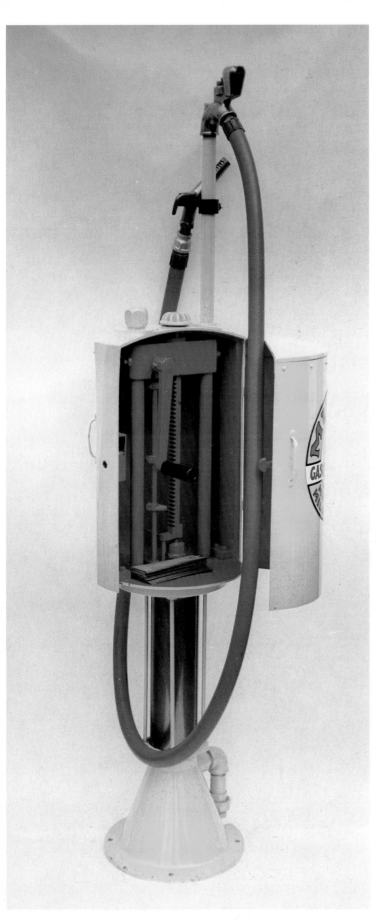

Milwaukee was the manufacturer of this circa 1915 curbside pump. This pump pre-dates visible pumps and was used in its later years to dispense kerosene. *Pump restoration by Gene Sonnen.* $1,200

Filtered gasoline was offered in this pre-1920s Wayne barrel pump. The dial and mechanism were exposed by lowering the shroud seen at the bottom. *Restoration by Gene Sonnen.* $2,500

Tokheim produced this deco-style pump in the late 1930s. It incorporated a spinner with a small visible glass bubble that the fuel would pass through. *Restoration by Gene Sonnen.* $2,600

Almost all of these 1950s style Tokheim pumps were produced with baked enamel as their outside coating. However, the one pictured here actually has a porcelain enamel coating. *Original unrestored condition.* $1,200

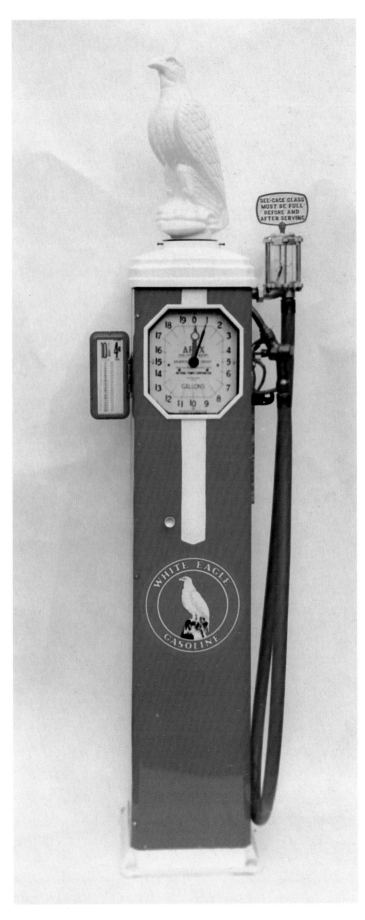

Apex produced this pump in the 1930s era. It utilized a see cage glass on its upper body. The dial design has come to be known in collectors' circles as a clock face. *Restoration by Gene Sonnen.* $3,400

Bennett was the manufacturer of this 1930s restored Koolmotor pump with pricing box. It also used a see cage glass. No doubt the see cage helped give visual confirmation that the motorist wasn't filling up with anything except gasoline. Note the Barq's Root Beer advertisement on the face of the pump. *Restoration by Gene Sonnen.* $2,800

Most early era service stations needed a way to fill oil containers. The device pictured here pumped oil from its lower storage compartment up through the mechanism and out to a waiting container. These are known as lubesters. Notice C.S. Oils is embossed on its pumping handle. $450

Crystalite kerosene was dispensed from this early hand-operated pump with banana nozzle around 1920. $300

Here's a top-side photo of the Bowser pump showing not only the dial detail, but its original tag.

Bowser was the manufacturer of this one-gallon nickel-plated kerosene pump dating to pre-1920. $425

Tokheim manufactured this pre-1910 pump to dispense gasoline. Early service stations often dispensed into cans or pails which were then used to funnel into the tank of automobiles. This dangerous practice was short-lived and soon was replaced by safer direct fueling methods. $500

Cedar Rapids, Iowa, is prominently visible around the pump's base. Also notice the patent dates of 1900 and 1903.

The two visible style pumps shown here are actually countertop mechanical dispensers for lighter fluid. These could be found not only at gas stations, but at truck stops, the local drugstore, and many other locations. They are included here because of their desirability in the petroleum collectibles market. $1,500 and $1,000

The 1940s saw production of this child's Shell service station complete with automobile hoist and fueling island. The toy pumps are representative of Model 60 Waynes and have the appropriate Shell Regular and Super Shell with Ethyl globes attached. $650

Don Larson of Hucksley, Iowa, built a limited number of
these unbelievable model service stations, taking over 200
hours to produce each one. They are unsurpassed in attention
to detail. This station has miniature Tokheim Model 36
Computer Pumps with operating handles, brass nozzles,
rubber hoses, light-up ad glass and globes. The pumps are
sitting on a solid cement platform. What service station
would be complete without a clock-face air pump or Bowser
kerosene pumps? Don's model contains both with light-up
globes. The roofing is cedar shingled with a cement chimney.
The lightning rod system includes white milk glass balls and
is appropriately grounded down the side of the building.
What a work of art! If this can't impress you, nothing will.
$2,200

PARTING SHOTS

#1 - This outstanding array of Shell petroleum products adorns the display room of Tom and Susan Dahl's "in house" Shell Station.

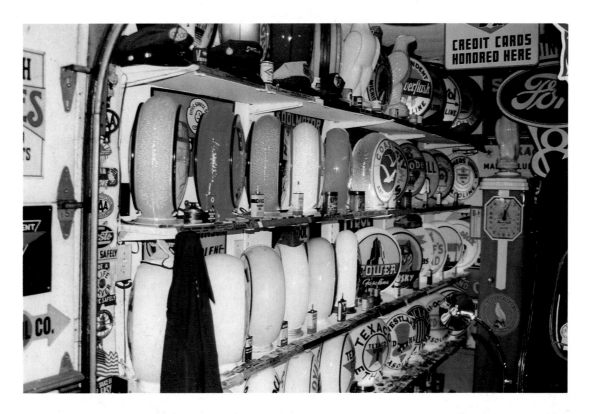

#2 - A few of the many beautiful pump globes enjoyed by Gene and Red Sonnen are visible in this wall shot.

#3 - This jungle of great stuff creates the desired effect in the display room for Gene and Red Sonnen.

#4- If signs are your thing, you'd be right at home here. Despite the limited wall space, Tom and Susan Dahl know there's always room for one more.